THE COMMON PL

The Common Place
The Ordinary Experience of Housing

PETER KING
De Montfort University, UK

Routledge
Taylor & Francis Group

LONDON AND NEW YORK

First published 2005 by Ashgate Publishing

Reissued 2018 by Routledge
2 Park Square, Milton Park, Abingdon, Oxon, OX14 4RN
711 Third Avenue, New York, NY 10017, USA

Routledge is an imprint of the Taylor & Francis Group, an informa business

First issued in paperback 2018

A Library of Congress record exists under LC control number: 2005011790

Notice:
Product or corporate names may be trademarks or registered trademarks, and are used only for identification and explanation without intent to infringe.

Publisher's Note
The publisher has gone to great lengths to ensure the quality of this reprint but points out that some imperfections in the original copies may be apparent.

Disclaimer
The publisher has made every effort to trace copyright holders and welcomes correspondence from those they have been unable to contact.

ISBN 13: 978-0-815-39757-1 (hbk)
ISBN 13: 978-1-138-62087-2 (pbk)
ISBN 13: 978-1-351-14740-8 (ebk)

Contents

Preface

My ambition here has been to write what I have termed a root book. Indeed the metaphor of the tree root plays a large part in what follows. Those rash enough to have delved into the quagmire of post-structuralist thought will doubtless appreciate the reference. A root book is 'the classical book, as noble, signifying, and subjective organic interiority' (Deleuze and Guattari, 1988, p. 5). It is the sort of book that tries to say something in the belief that saying something matters. Of course, this is hopelessly naïve. As post-structuralists such as Gilles Deleuze and Felix Guattari have tried to teach us, messages are not straightforward things, but are inherently tied up with what they exclude, with what they seek to inhibit. Thus to write a root book is to create a textual silence that, as it were, screams out to us. A root book, so the post-structuralists would have it, is representative of hierarchical, rigid monocultures which deny the flux and contingency that theorists like Deleuze and Guattari claim to see all around them.

Yet what I consider to be most important about housing is that, when it works well, it is the precise opposite of flux and contingency (and when it works badly it is because there is too much flux). Our housing is one of the things that brings stability to our lives, by locating us, settling us, and by giving us a place. What is more this place is something we can take for granted as it is the background to our lives. Housing, I wish to suggest, is something that is ordinary, in the sense that it offers us no surprises and no shocks.

Accordingly, I can readily admit to the 'faults' that Deleuze and Guattari see in the 'arborescent system': or rather, I see them as virtues and not as faults and, in consequence, I seek to consider fully just what it means to be rooted. What I wish to stress in this book is the sense in which we are linked into something. We are literally grounded: we are stuck in the ground that nourishes and nurtures us. And furthermore, this nurturing is clear to us and based on a direct and obvious link, just as it is obvious which part of the tree is stuck in the ground.

This might lead the attentive reader to believe that this book is about traditions and conventions, and is therefore somewhat conservative. I would plead guilty on all three counts. However, it is about traditions and conventions only in a general sense: it does not dwell on any particular traditions or conventions; and the book is conservative in a dispositional rather than an overtly political sense (insofar as the two can be separated). It is an example of what Anthony Giddens (1994) has defined as 'philosophical conservatism', being a general predilection towards the known and the tested and against the disruptive nature of radical change. Much of what is good in conservative thinkers, such as Edmund Burke (1999), Michael Oakeshott (1991) and Roger Scruton (2000, 2001), is not that they are opposed to change *per se*, but rather that they are asking us to consider the consequences of change. These thinkers want us to acknowledge that change is often irreversible, and moreover, often necessitates a step into the unknown. Accordingly, they ask us to consider just what it is that makes us want to change, and what it is that is so bad about what we now enjoy. Their conclusion, if one can generalise about thinkers who are so deliberately unsystematic, is that change is not merely acceptable but is *necessary* where it can be seen to strengthen those institutions, practices and relations which nurture us. This book, in its more narrow and shallow way, is trying to make the same sort of argument. I am seeking to encourage those who make a living 'out of' housing to consider why they want to change things, and to reflect a little more on what it is they are seeking to transform. An inevitable part of this is to consider just what it is that housing is for, and why we should seek to protect it *as it is*.

My method in undertaking this study is impressionistic, and there is a particular purpose in this choice of method. This book is not attempting to be different for its own sake, or deliberately to oppose the mainstream. My motive in choosing impressionistic approaches is that I see housing in these terms. As I explain in the book, I am not talking here about the sort of housing that comes in aggregates, that is made by policy and needs government subsidies. What I am referring to is the on-going activity that frames our daily lives, the protected intimacy of caring, sharing and mutual security (Bachelard, 1969; King, 2004b). We do not experience our housing in the form of coherent arguments, rebuttals and discourse, but through association and the linkage of personal actions to specific places at particular times: we are located through the specificity of enclosed space and the immediacy of personal relations. What we should seek to do then – as I have tried to do in what follows – is to capture this sense of housing as a series of impressions. It seems to me that one way of doing this – but perhaps not the only way – is through the use of film, anecdotes and personal reflection.

The book might therefore be said to lack the full panoply of academic equipment which could have been brought to bear here. For instance, I could have undertaken a critique of the sociology of everyday life, or of material culture, or have engaged with thinkers such as Jean Baudrillard, Pierre Bourdieu, Edmund Husserl or Maurice Merleau-Ponty, who have all had things to say about what it is to be ordinary. But to have done this would have created a number of problems. First, it would have made it a very long work. Second, it would have been a fundamentally different book, in that it would have taken off at tangents that would not have helped in developing my own point of view. So there is a greater purity, if that is the right term, in this approach. Third, what I did not wish to do was to restrict myself merely to a consideration of what others have said about the ordinary. Of course, I have made use of the work of others, particularly Stanley Cavell and Stanley Rosen. But I hope to have developed their concepts and ideas and made something new and different. What worries me about the emphasis on 'the literature' is that it stifles thought, and also lets critics off the hook: instead of having to engage with a particular position, having to understand it and make sense of it, it is seen as perfectly acceptable to ask 'where's Merleau-Ponty?' and to assume you have demolished that position. One ought rather to engage with the argument as presented, and seek to understand the terms upon which it is written, before running back to the security blanket of 'the literature'. If no one is prepared to do this then one wonders how there will ever be any new ideas generated in housing studies.

My fourth, and most important, reason for taking this approach is that I want my discussion to come out of the activity of housing itself, to flow from the manner in which we use our housing. I want to build up a conceptualisation of housing from the bottom up, using materials that relate directly to housing. Having done this conceptualisation, we can then reach out for commonalities between, and support from, other concepts and theorists. This does not mean, of course, that we ignore existing theories and concepts as we build our own – we do not start with a blank slate – but what I insist is that we start with housing.

Hence I do not wish to attach housing to a theory. Instead what I am seeking to do, in all my naïveté, is to create concepts *of* housing, and by extension to encourage others to create theories *of* housing. And I believe that we can do this and ought to, because we can glean particular insights that take us beyond the materialist conceptions of housing that tend to dominate the field. What I am seeking is to show that housing itself can be the very basis for theorising, rather than being just the 'case study' which justifies or refutes a particular theory.

I have, I am afraid to say, something of a passion for trees – not to the extent, you understand, of wanting to hug them, nor to sit up in one until the bulldozers go away. However, I do like looking at them and taking photographs of them. Accordingly, I have taken the tree – particularly its rootedness and organic wholeness – as a key image in this book, and consequently I have included a number of my photographs of trees, and parts of trees, in this book, as well as some other images. These photographs are intended to add something to the discussion, to set a mood or tenor for the ensuing discourse. The particular images at the start of each chapter aim to resonate with a particular sensibility or motif within that chapter. They seek to offer some clues rather than to 'explain' anything of themselves. What I hope is that readers will see the photographs in the same way that the score adds something to a film. What I have not chosen to do is include any stills from the films I discuss. Whilst some might have seen this as helpful, I do not, any more than I would include a photograph of page 5 of *A Thousand Plateaus* to illustrate the quote used in my opening paragraph. What I hope I will have done is to encourage readers to return to these films or to see them for the first time.

Whilst writing is a solitary process, one still of necessity comes to rely on many others to get one's thoughts between two covers. I therefore have many people to thank. First, I am extremely grateful to my colleagues at the Centre for Comparative Housing Research, who continue to provide a supportive and fertile environment in which to write and think. In particular, I am grateful to CCHR's director, Tim Brown, both for his perceptive comments on my work over the years, but also for taking on many burdens himself so that others have time to write and do research. I sometimes think that the word 'indispensable' was invented with him in mind. Both Nick Mills and Jo Richardson read an earlier version of the manuscript and I am grateful for their comments and encouragement. Valerie Rose has been a very supportive and enthusiastic editor at Ashgate, who has shown considerable confidence in this rather eccentric project. I would also like to thank Matthew Carmona for agreeing to include my book in his fine Design and Built Environment series, and to those anonymous reviewers who also supported the project.

As usual I have relied greatly on my family. My brother, Graham, has again helped with the scanning and preparation of the photographs. His patience and good humour with his technophobic elder brother is unflagging. Finally, I am so indebted to those whose common place is also mine: to my two daughters, Helen and Rachel (despite the fact that they have insisted on being in another of my books!); and to my wife, Barbara, for proofreading and excising the more wilful examples of my naïveté, and not least, for continuing to put up with me. What more can I say?

Peter King
March 2005

Introduction

Much of what constitutes our experience of housing is quite ordinary. Our housing is a familiar space, full of familiar things. It is the one place where we seek to avoid the exceptional and the surprising. It is commonplace, even nondescript, and fits around our regular routines so that we lose sight of its complexity. This is both ironic and important: ironic because, whilst policy makers and academics seek to attain transformation or 'step changes' in housing, what we seek from our own housing is stability, routine and a means to avoid change; and it is important precisely because we need this stability and lack of change. Our personal experiences of housing are significantly different from the manner in which we discuss housing policy or housing markets. The manner in which housing is used differs markedly from the manner in which it is created, allocated and accessed.

Housing, it is both needless and essential to say, is something we live in. It does not consist of policy documents, strategic plans or Best Value inspections. Housing policy concerns itself with the conscious – with plans, strategies and regulations. Yet this is not how we live: we do not live by committee. Policy has no connection with what we want in our housing and with how we use it. Housing policy is thus something completely different from the *activity* of housing and needs to be kept separate.

Indeed this concern for the conscious plan is inimical to the use of housing. We can see this when we consider the idea of the 'step change' as proposed by the Blair government in the UK (ODPM, 2003). This is an attempt to transform housing policy, with the implication that we move from the rather sleepy and slow policy development of the recent past onto a higher and faster Third Way plane. It is intended as a radical transformation, a considerable progression, a move or shift on. Yet, regardless of the plan's success or failure, why would we want this in our housing? Why do we want to be moved and shifted on? What is so wrong about how we live that it needs to be transformed according to some national plan? What are we doing that is so in need of change?

I would suggest that the virtue of our housing is that it protects us from change: it allows us to stay put, and this is important because that is precisely what we want to do. We wish to maintain what we have, or else to build on it and develop it on our terms and in our time. We seek to be free from intrusion, especially from large impersonal forces that we cannot control or understand.

The basic dichotomy here is between innovation and the ordinary: between the desire to change and the need for stability. The activity of housing is ubiquitous: it is something that we all have and use (or seek to, and would see ourselves as being seriously harmed were we not to have it). It is something that is always there and ready for us and, in consequence, we would try to ensure that change and transformation are minimised in this private sphere of our lives (King, 2004b). The problem is that things like housing that are so common will tend to lose our interest:

there is no novelty, no opportunity for change and transformation. This ubiquitous activity of housing, from a professional perspective, appears boring. Johnston Birchall (1988) makes this point when he criticises architects for trying to innovate:

> It is the sheer boredom of the subject that has propelled architectural students to those absurd flights of fancy, which combined with successive attempts to adapt a simple but labour-intensive technology to the assembly line have provided us with most of our housing problems (p. 15).

Perhaps this is a comment we should remember when the apparent 'step change' promised in English housing provision is to be achieved in part through the use of off-site (assembly line?) techniques which are generally considered to be more expensive than traditional methods, as well as being relatively untested. Birchall was concerned with the architectural and planning failures of the 1960s and 1970s. But it is a quote that we might start to apply again, especially when housing associations such as Peabody Trust are criticised by the Audit Commission for being more concerned with innovative schemes such as the 'award-winning BedZED scheme in Sutton, North London, (which) cost £10m more than had been planned' (*Housing Today*, 14 May 2004, p. 9) than modernising their older properties. The problem with 'step changes', we might say, is that one often does not know where one's feet will land.

But this is not a book about housing policy and there is little in the way of policy discussion in the rest of the book (except for some general criticisms in Chapter 4 and the Conclusion). Government housing policies are about production and consumption, and are based on a material conception of housing, which sees things rather than activities and meanings. Housing policy has nothing to say about use, about what we do once the front door is closed. Yet it is the use of our dwellings that concerns us most and for most of the time: this is what the reality of housing is for us, not policy statements, strategies and 'step changes'.

So my aim is to discuss the opposite of the innovative and the transformative. This book is about housing as something ordinary and common. The discussion is largely conceptual and not driven by the immediate concerns of policy or specific housing issues. I want to go beyond time-bound policy discussions that concentrate on specific measures and instruments. These policies will change as governments and intellectual fashions come and go. But what remains is the nature of housing as an activity, as something we use and take pleasure in. What this book tries to consider, then, is not housing as policy but as a cognitive process.

I want to deal with what the ordinary is, how it is manifested, and, of course, how it links with housing. The ordinary is neither a thing nor a theory, but a disposition, and hence it is about the epistemic conditions between beings and things. It is a phenomenological adherence to our conceptualising of things known to us. The ordinary is about creating sense of our world and how we are able to use it to root us in the world. It is about the sensing of both ubiquity – that these things are all around and seen by all as such – and specialness – there are certain things that are just ours and there for us.

The ordinary can be seen as circular, as reiterative. It is constantly bringing us back to the things that we know, that we love, that we care for and share. We are always in the midst of the ordinary. Without this reiteration we would not be comfortable or secure. We need the certainty of known things in order to be placed and then venture out into the world. We do not wish to change our step, but to keep a confident tread based on the known routes out of a place we know as *ours*.

The method of this book too is circular and reiterative. What I have attempted is an immersion in the ordinary, where we are constantly amidst the ideas and arguments, so that even as the ordinary becomes apparent, we see ourselves as being part of it. This is instead of seeing it as an argument to propose and defend, or as a position one can take until it is disproved or falls apart. We do not, properly speaking, present an argument for the ordinary, but rather we just try to see it, to see the frame that surrounds us. So in this book there is no attempt to develop something. If there is a journey it is along known paths with a view to coming home safely. My idea is not that we should arrive at somewhere new: this would involve leaving what we have and accepting that the purpose of housing discourse is to transform ourselves and our environment. Housing discourse, like all intellectual reflection, is to help us come to terms with and to understand what we already know. In accordance with this, what we have in this book is a discussion that circulates, that travels around, picking up ideas that appear useful and interesting to us and putting them together with others, in the knowledge – or perhaps, the hope – that the journey will conclude somewhere we find congenial.

But is there any point to a journey when we know the route and that we can get back safely? This may seem to run against the spirit of academic endeavour. This is not the model of so called 'blue skies' thinking where we seek knowledge for its own sake, regardless of any utility it might have. Instead we appear to have our eyes trained to the floor picking out the safe route. Yet, when we know the way we are on, say, a familiar walk, we are able to look about us, safe in the knowledge of where we are, and that we are able to pick out the changing seasons, the things that we might not have noticed before amongst the more familiar. And we begin to see more deeply, even if we do not see more of anything. We start to see complexity in the familiar, to see beauty and to revel in it; and we revel in this precisely because this is so close to home. This is not exotic beauty or remarkable because of its strangeness, but because it just is there and open to us. What is remarkable is that we can now notice it, that we have become opened up to what encloses us and which we see as tools, as taken-for-granted parts of our lives.

Using Film

This book, then, is very much about looking and seeing. I describe the ordinary as a frame, which we quite often fail to see as we regard the main picture it encloses. We have to try hard to see the frame, and what it does for the picture. I also choose to refer to housing as a stage, as the background to our activities. These are important metaphors in coming to an understanding of the ordinary. But they also point towards the appropriate method for such a study. Housing is something that is personal to us, even as we know it is a common and shared experience. Things that are so personal, and which frame our actions, are commonly experienced impressionistically. We gain only glimpses of their full significance: we may on occasions stop to reflect on where we are or on what we have done; in times of crisis or trauma we may be forced to bear down on important places and times in our lives; major shifts in our lives make us look back at ourselves. In all these situations we may recall places, faces and things that have been significant and which come out of the sheer commonality of our ordinary experience. But there is no consistent narrative, and no logical development: there are merely scenes, played before a consistent background.

Such an understanding of the manner in which we relate to our housing suggests some care with the methods of investigation. In particular, what means have we to understand this impressionistic sense of recall we have? It seems to me that this approach needs to make recourse to sources capable of capturing this impressionistic sense. The crucial elements here are time and space: our impressions are time-bound, yet they also connect to a particular space or place; they are located to the specific. What we remember are specific spaces at a particular given time. This notion has been enunciated by the Russian literary critic, Mikhail Bakhtin, in the concept of the *chronotope*. He defines this concept (which literally means 'time-space') as 'the intrinsic connectedness of temporal and spatial relationships that are artistically expressed in literature' (Bakhtin, 1981, p. 84). The artist aims to connect time and space into one, and where it works well, 'Time, as it were, thickens, takes on flesh, becomes artistically visible; likewise, space becomes charged and responsive to the movements of time, plot and history' (p. 84). What this meant for Bakhtin was that questions of the self could only be dealt with when seen as questions of specific location (Synessios, 2001). Our sense of self, or our history, is also our sense of time and place, and this is always specific.

Bakhtin's concept of the chronotope is useful and one that need not be restricted to literature. As Natasha Synessios (2001) has suggested, this concept was influential on the work of the Russian film maker, Andrei Tarkovsky and, in particular, his film *The Mirror* (1974). This film (which I discuss in Chapter 5, along with several other films of Tarkovsky) is concerned specifically with the locating of memory, of remembering the places of childhood. This shows that we can extend Bakhtin's concept beyond literature and to other areas of artistic endeavour. Indeed one might suggest that film is particularly suitable for this fusing of space and time, with its attempt, as Tarkovsky himself saw it, to stop time. Film is an artistic medium specifically intended to hold up time and to create impressions using time and space. I have therefore used film as one of my main sources, in addition to personal impressions and general critique.

Housing studies, as an academic field of study, has been largely immune to the influence that film has had on popular culture. Much housing research relates back to the notions of social administration and political economy that have dominated British social science since the 1960s (Clapham, 2002). However, within the more general field of urban studies, a number of academics have used film, narrative and fiction in order to elucidate their arguments. Peter Dickens (1990), for instance, draws on the fiction of Martin Amis, William Boyd and Tom

Wolfe to show particular aspects of the urban within the milieu of the Reaganite-Thatcherite 1980s. Another example of the use of fiction is where Roger Burrows (1997a, 1997b) talks of the cyberpunk fiction of William Gibson and Neal Stephenson as exemplars of a sense of the urban as polarised space, as divisive and divided. In this sense, these novelists can be said to be prefiguring urban social theory, even though they are looking forwards rather than backwards in the manner of Bakhtin.

In terms of film, Mark Shiel and Tony Fitzmaurice (2001) have attempted to make connections between film theory and key urban theorists such as Manuel Castells, David Harvey and Henri Lefebvre. Katherine Shonfield (2000) has used a range of films, from Roman Polanski's *Repulsion* (1965) and *Rosemary's Baby* (1968) to the Ealing comedy, *Passport to Pimlico* (1949, directed by Henry Cornelius) to explore the architecture of cities. Specific to housing, I have used films such as David Fincher's *Panic Room* (2001) and Krzysztof Kieślowski's *Three Colours: Blue* (1993) to consider issues such as security, anxiety and loss (King, 2004a, 2004b). This work has shown the possibilities of using film to elucidate our experience of housing. Film has the virtue of always relating us to specific individuals in a given time and place, and thus opens up the opportunity to explore this relationship. If we see housing as an activity that involves the day-to-day relationship with significant others and a loved place, then we can usefully use what we might call the *chronotopicality* of film.

Housing is ubiquitous, as is film. Film, we might suggest, is a measure of a popular culture that is now so pervasive. This link to the notion of the 'popular' is important and significant. It can help us connect the academic study of housing to the level, content and sense at which most people enjoy their housing – as a place where they can be comfortable, relax and seek to enjoy themselves, and film is a key form of relaxation and enjoyment. If housing is so ubiquitous an activity then we should expect to see its traces everywhere and therefore be able to recount them. Housing has been the background to many narratives, be it the homelessness of Mary and Joseph; the gingerbread house tempting Hansel and Gretel; the straw, wood and brick houses of the Three Little Pigs; or of films such as those discussed in this book. Housing is the background to much of what we do and so descriptions of it are imbued with significance. As it is so ubiquitous we should expect it to be used in fictions, be it positively or negatively, centrally or marginally. We should also try to engage with these fictions to try to determine why it is that this particular 'intrinsic connectedness' has such as resonance for us.

This might suggest that housing is reduced to a banality, or a mere token in the background. Yet we use symbols all the time in our discussions of housing. We only need to think of examples such as the 'suburban semi' to connect with certain ideological tropes. We should not therefore be surprised to see housing used symbolically in other ways which convey meanings, set up a scene or create a mood. I hope by the end of this book to have demonstrated that stating that housing is the background is its great strength and not its diminution.

But there might be another objection to my discussion here, along the lines of 'where's the housing?' Several chapters in the book are only tangentially 'on' housing. A more relevant term might be dwelling in its fullest sense of 'settling on the earth'. The discussion is often more about the general, and relates to place, space and home as generic concepts. Yet what this objection really means is '*where is the discussion of housing policy, and where is the consideration of the role of government, specific policy measures, and subsidy systems?*' However, the purpose of this book is precisely to go beyond these narrow discussions and instead to try to gain a fuller appreciation of the significance of housing *in the general sense*. So one needs to respond to this question by asking why these critics are so narrow in their ambitions, and why we should seek to contract the concept of housing to something so shallow. When there are so many avenues we can pursue, be they in philosophy, social theory, literature and film studies, why do we seek to restrict ourselves to narrow policy discourses? There is a perversity in the search for relevance, and the consequent attachment to all things policy driven, and this is that there is little intellectual development in the field. By concentrating on housing policy we allow our studies to be driven by others, be they governments or those funding our research and seeking to use our findings. As a result there is little opportunity to develop the concepts of the field and substantially to advance our understanding. In my view, if we take a step away from policy, and gain some distance, we might be better able to develop intellectually and hence challenge the drive for relevance and yet more relevance that actually stops us thinking.

The book is presented as a series of expositions of the ordinary and how it relates to housing. Chapter 1 considers some of the key concepts of the book: roots and ruts; acceptance and accommodation. I also set myself up in opposition to post-structuralist thought, with an extended critique of the dichotomy of root and rhizome. I use the film *Spring and Port Wine* to introduce the key concept of *accommodation*, which I develop throughout the book. The ordinary is properly introduced in Chapter 2, where I consider the fascination of the ordinary and why it is so important. This is followed in Chapter 3 with a discussion of the properties of the ordinary as developed in the work of Stanley Rosen and Stanley Cavell. In this chapter I also consider how we might discuss the ordinary and how it relates to housing. Accordingly, I return to the idea of the chronotope, briefly discussed above. Chapters 2 and 3 move from the ordinary as unseen towards a fuller recognition of what we are in the midst of. However, I make no attempt to define the ordinary, or to construct a theory of the ordinary. This is because the ordinary is a disposition rather than something that can be constructed as a testable theory. In consequence, we need to rely on impressions rather than argument, and this is why I choose films as my main source for considering aspects of the ordinary in the later chapters.

Chapter 4 considers the idea that housing is the background to our lives, which allows us to pursue our ends. I demonstrate this through a discussion of films ranging from well-known Hollywood productions such as *The Matrix* and *Unbreakable* to important European films such as *Summer with Monika* and *Ordet.* This leads to a consideration of the epistemic conditions of housing and how they link to the properties of the ordinary. This is contrasted with the limited vision of housing policy which barely scratches the surface of dwelling in its full extent.

In Chapter 5, I consider how we still maintain ourselves when we are 'out of the ordinary' in a physical sense. My aim here, using the films of Andrei

Tarkovsky, Bela Tarr and Gus Van Sant, is to show the importance of remembrance of favoured places when we are displaced or exiled, and where the limits of the ordinary's capabilities may lie. This allows me to demonstrate two things: first, that the ordinary is not merely a physical state, but also relies on memory; and second, that there are limits to the ordinary and thus some places exist where we cannot find acceptance, both in the physical and emotional senses. Chapter 6 considers attempts to make our housing extraordinary though design. Using the film *Mon Oncle* as my starting point, I show how this is self-defeating, in that design can only succeed by threatening the private and personal nature of our ordinary experience. But, in the face of this onslaught on the ordinary, I show how resilient it is, through the manner in which we assimilate technology into our daily lives.

My aim is to return full circle in that, having explored the nature of the ordinary and its relation to housing, I shall return to the roots of my discussion. This means that I see my discourse as being rooted into a particular view of the world, a way of seeing, a disposition. It is my hope and my intention that the approach I have taken in this book will of itself help to demonstrate the virtue of the ordinary. I seek to make clearer what is already around us, and show how we should revel in it rather than seeking always to make ourselves anew. I want to show how we are rooted in a place and how we have a limited number of known and trusted ruts by which we feel able to connect with the wider world. My starting point on this journey is therefore to show how we are stuck in our place by the roots.

Chapter 1

Roots and Ruts

This is a root book. It is one that is grounded in something solid. It is secure in its antecedents, from where it takes its sustenance. It buries itself in something substantive and sustaining. There is a certainty, a straightforwardness, and this is comforting. As a root it holds secure, gripping tenaciously to the surrounding material, without which it has no future, no prospect other than a wilting death.

I use the term *root* consciously, and seek to use all its registers and resonances. To say we are rooted is to suggest we are connected, that we are placed. As Roger Scruton (2004) suggests, it is an awful cliché to say that 'we are rooted in the soil', but then, to coin another one, clichés become clichés because they have some truth in them. We have a sense of place and we feel located. This may be a small space, like a village or even our own dwelling, or it may be a community or a nation (Weil, 1952). But we can identify with that place. We are rooted into that place. This is another way of suggesting that we gain meaning from what is around us (King, 2004b).

Yet this sense of rootedness is not merely about meaning. It is also about obligation. As Simone Weil (1952) suggested, being in a community – being tied to place – brings with it an obligation. We are committed and this is a two-way process. We gain from our connection with others and with an ideal, but that also means that we too must work to protect and to nurture that connection and that ideal. Being rooted in the soil brings with it an obligation to maintain it, to sustain it and to put back more than we have taken out. Our use of the soil is a form of trust (Scruton, 2000), held by us for those gone and those yet to come. But it is also about the way we can trust our surroundings, in their regularities, their assuring qualities and their certainty. This place then is very much what we are stuck in: it is a rut, *our rut*.

But there is more to 'place' than that. Places are often substantial things that can stand up to the elements, that can resist their buffeting. Our dwelling covers us and protects us. It allows us to be intimate with those we love and share with; it helps us to feel secure by offering us privacy (King, 2004b). But to do this it must be grounded. A house is built on foundations. These foundations go down into the ground and remain there as solid embodiments of our need for roots. They do not sprout, link up with other foundations, create networks or seek out difference. They stay where they are, and for this we are grateful. Foundations are anchors; they hold the house tight to the ground. They form the sustaining link with a place. Dare we say that foundations root the house into the ground?

But, of course, the dwelling is itself the manifestation of our rootedness. It is the fruit of our desire for place. When we put our dwelling in *that* place we ground it in significance. The roots, we are tempted to say, make the place for us. Their groundedness brings the place into our senses; they give meaning to place.

We put down roots, but this is not always or ever of our own choosing. Often it is where we fell and 'took' in the ground. We are located here for reasons that are unclear to us, even as we realise the importance of our being here. But we need not understand our rootedness for it to work for us. We have fallen already on made ground, that has its ruts carved by those who preceded us with their traffic and commerce. So, as we put down roots, we find that there are ruts (routes) already leading to us. Indeed, as Roger Scruton (2004) reminds us, it is often only because we are within well-trodden ruts – that we can say our roots go back generations – that we will be accepted in a place and feel able, and are allowed, to call it *ours*. It is only once we have shown our obligations and commitment to a place that our roots are acknowledged. This is often important, even though it might be considered exclusionary and divisive. It is important precisely because it will guard us against false idols and the pernicious belief that we can take up our roots and walk.

The root is connected to enlightenment: it holds the plant up, so that it can reach up to the light. It takes its energy from both above and below. The root is a source of nutrition from close by. It is within the soil, taking up sustenance from around it. But it also supports the plant as it takes its energy from a source that is distant, seemingly eternal and certainly universal. So do we link into that which is near: we take the shelter offered by our rooted dwelling, but we are nurtured also by more distant, impersonal and long-lived entities like a community and traditions and customs. The house provides this for us as it offers the secure footing that allows us to explore outwards. Christian Norberg-Schulz (1985) used the analogy of axes and pathways, as the means by which we are oriented in our environment and our community. These are the ways by which the meaning of our surroundings is signified to us. Instead of using these terms, however, I wish to talk about roots and ruts. The similarities are clear: both axes and roots are fixed points and ruts and pathways are marked routes, which allow us to explore what is around us.

Yet I wish to be more restrictive than Norberg-Schulz. I wish to emphasise the fact that we are often stuck where we are, and to move we would have to be dug out. Likewise, we cannot go where we would like, but only where we are directed. There are often only a very few ways we can go. And also we cannot use the excuse that the pathway is not clear: it is not an indistinct mark across the landscape, but often a deep rut presenting us with the one and only way.

This rather restrictive position can be seen as a corrective to the emphasis that post-structuralists place on flux and contingency. I want to show that post-structuralism is mistaken in this emphasis and that it misunderstands human nature, particularly the sense of humans as dwellers, as people settled on the earth. This idea of settling is inimical to flux and contingency, which are states in which no one can feel secure in their place and take comfort and complacency from it (King, 2004b). My particular target here is the discussion of tree root and rhizome that forms both the method and structure of Gilles Deleuze and Felix Guattari's seminal book, *A Thousand Plateaus* (1988). The rhizome is emblematic of contingency, flux, relativism, of all that post-structuralism is meant to be, but which our dwelling is not and cannot be. To suggest that I am presenting a root book is thus to oppose this influential example of post-structuralist thought and all it stands for.

In undertaking this critique of Deleuze and Guattari it will also become clear just what I mean by a *root book*. I want to establish my targets, but also to be more positive and to lay the platform for the discussions that follow. What I wish to do is to make clear the route I intend to take – to show what rut I am in – and to show what sort of argument I am rooted in. What I am attempting is a distinctive description of housing, one based on the notion of ordinariness, of roots and ruts, and the means by which we stay put and pass through our surroundings with a sense that we are accepted and can, in turn, accept what we have and where we are.

The Root and the Rhizome

Gilles Deleuze and Felix Guattari collaborated on a number of projects in the 1970s and 1980s, the most influential of which are two books loosely connected by the subtitle *Capitalism and Schizophrenia*. These two books, *Anti-Oedipus* (1983) and *A Thousand Plateaus* (1988), are complex, (sometimes) amusing and perplexing books. In the foreword to *A Thousand Plateaus* the authors say that each chapter can be seen as a plateau that can be read independently of the others and in any order. Indeed, apart from the conclusion, it is hard to see what order the book has. It is rather a rambling, complex and very difficult book, which leaves one with a series of general impressions as well as some particular insights on Marxism, psychoanalysis and capitalism. It is though almost impossible to state what the book is about or summarise it succinctly in the manner one might do with other works of philosophy. Both these books are inherently anti-humanist with their constant references to the analogy of the machine as a key concept, but there is no system to be derived from reading Deleuze and Guattari.

But this, of course, is precisely as they would have it, for the complex structure of *A Thousand Plateaus* is as important as the particular content of the work. In this sense it has become emblematic of post-structuralism. As well as representing the anti-humanism and materialism central to a critique of structuralism and phenomenology, the book also presents a method and approach to discourse, that are themselves post-structuralist. This is shown by the key dichotomy, presented in the opening chapter of the book, between the tree root and the rhizome. This discussion has become important in that it can be seen as a key statement of the post-structuralist method, which emphasises what has become so significant a development out of post-structuralism. This development is that of *difference*, of the belief that there is a multiplicity of possibilities open to us rather than just one or two routes. Instead of seeking unity, boundedness, a fixed point, we should rather see networks with no core or centre. We should see multiplicity of form rather than a definitive structure. *A Thousand Plateaus* takes on this structure itself and is thus an example of what post-structuralism is, as well as presenting a key statement of method and outlook. Another way of describing difference is to see it as *extraordinary*: post-structuralism, then, can be seen as the perpetual search for the extraordinary.

What I wish to do is to explore this key dichotomy of the root and the rhizome for the light it can shed both on method and on more substantive issues. I wish to use it as a means of presenting my own preferred method, and also some of the key principles which underline my argument on housing and dwelling. But first I wish to deal with a possible objection. In the housing literature Deleuze and Guattari do not feature heavily. Indeed much of the controversy about post-structuralism seems to have passed housing studies by. It would be fair to say that I am not criticising Deleuze and Guattari out of any sense that they are dominating housing discourse. If I wished to pick a target then the ideas of Michel Foucault would be more pertinent, particularly with his ideas on discourse, governmentality and the gaze. But even here, it would be difficult to claim that Foucault had taken the world of housing research by storm.

Why I believe Deleuze and Guattari are important, therefore, is not because they are influential in themselves, but rather because they demonstrate a key element of post-structural thought that has pervaded much more widely than other more abstract concepts. This is this notion of difference or the extraordinary that I have already identified. It is this stressing of the outside, of the other, of what is different that I wish to contrast with my discussions of the ordinary and change. This is not because I wish to denigrate minorities, nor do I wish to ignore the very real needs of certain groups who are, as it were, outside of the mainstream. What I am concerned about rather is that an over-concern for otherness, for pointing out how we differ, how diverse we are, and how we should concentrate on the particular, reduces the possibility for solidarity. This emphasis on difference denigrates what we hold in common, what is universal and what is ubiquitous in our surroundings.

I would argue also that a concern for difference has the very opposite effect of what post-structuralists call for. They see the emphasis on difference as an attack on power and hierarchies. Deleuze and Guattari see their rhizomic method as anti-hierarchical and anti-authoritarian. Yet, as I argue in *Private Dwelling* (2004b), the effect of post-structuralism is often to leave nothing in place but naked power. The attempt to rid notions of desire of all limits, to breach the hegemonic laws and proscriptions of a society, is to leave nothing to protect the weak. It does not liberate anyone but those already capable of acting for themselves. The privileging of otherness over order, of possibility and multiplicity over limits, is to allow those with power to exercise it without recourse to any communal reproach. Therefore the very attempt to be transgressive, to go beyond the limits of a hierarchical order, is to jeopardise those incapable of defending themselves. What post-structuralism is suggesting is that, in the name of difference and liberation, we tear up the roots that support and sustain us: because roots limit us, because they tie us down, we should be rid of them. But, of course, the consequences of doing so are to cut us away from the very thing that nurtures us.

There is then a very real danger in emphasising difference and transgression of the social order. In the terms of my later discussion, this danger comes about because we privilege the extraordinary over the ordinary, the uncommon over the common. Housing discourse concerns itself not with the ordinary or with what we have. Instead it is concerned with transformation (King, 2004b), with those that have nothing or not enough, with step changes in policy to bring about rapid change: it is overly concerned with what can be or ought to be, rather than what is. One can suggest, perhaps a little unkindly, that housing policy and discourse are concerned with finding new ways to create the old problems. This is because of the rush to transform, to be concerned with tomorrow and never today (and certainly not to learn from yesterday), to take us to the next level, even if it means destabilising our current place. A discussion on this desire for transformation and its roots in theory is therefore an important one. We need to understand what is meant by this desire for difference and why it is deemed so important. Only then can we properly contest it. Hence, even if Deleuze and Guattari did not build houses, we can learn something from what they said and what they meant.

So let us consider this distinction between root and rhizome. As portrayed by Deleuze and Guattari, the difference between the two notions is that of a limited determining structure and a loose network. A root is a thing 'which plots a point, fixes an order' (1988, p. 7). This is very different from a rhizome, which is varied and diverse: 'The rhizome itself assumes very different forms, from ramified surface extension in all directions to concretion into bulbs and tubers' (p. 7). Unlike roots which are seen as being singular in structure, or at least very limited, rhizomes can take a multiplicity of forms from the highly disaggregated to the dense.

Deleuze and Guattari suggest that the root concept is an attempt to reflect nature: root books are seen as tracings of nature, as simplified copies. They state that 'The tree is already the image of the world, or the root the image of the world-tree. This is the classical book, as noble, signifying, and subjective organic interiority' (p. 5). A root book is therefore one with a traditional structure which tries to impose an order on the world. But, Deleuze and Guattari claim that 'Nature doesn't work that way: in nature, roots are taproots with a more multiple, lateral, and circular system of ramification, rather than a dichotomous one. Thought lags behind nature' (p. 5). But they claim that even more sophisticated attempts at definition, where roots are not seen as being singular, are also doomed to failure. These attempts are still too simplistic, and Deleuze and Guattari want to reject the whole notion of trees and roots as a metaphor for discourse. For them, 'The tree and root inspire a sad image of thought that is forever imitating the multiple on the basis of a centred or segmented higher unity' (p. 16). By insisting upon unity the root metaphor cannot but fail to describe the multiplicity of possibility Deleuze and Guattari wish us to see as the reality.

They sum up the apparent problems of the root metaphor as follows:

> Arborescent systems are hierarchical systems with centres of significance and subjectification, central automata like organised memories. In the corresponding models, an element only receives information from a higher unit, and only receives a subjective affection along pre-established paths (p. 16).

In opposition to this they present the rhizome. Their claim is that in presenting this dichotomy they blow apart the whole notion of binary division. By presenting a view based on multiplicity they suggest they can dispense with simple dichotomies, and their claim is that the contrast between root and rhizome demonstrates this amply. Rhizomes are different from roots: they have no determinate form. Indeed Deleuze and Guattari go so far as to suggest that some animals can be rhizomic, for instance, rats or ants, who appear to operate in concert almost as collective entities, but without any form of central co-ordination. They also see burrows, 'in all of their functions of shelter, supply, movement, evasion, and breakout' (pp. 6-7), as a rhizome.

What are interesting are the particular facets of the burrow that they consider. They stress mobility and movement rather than stasis and stability, which might be seen to be the point of a dwelling (King, 2004b). Indeed the most popular fictional residents of burrows, hobbits, are portrayed by J.R.R. Tolkein as home-loving creatures of habit and routine, who dislike travel beyond their known surroundings: a hobbit, for one, would not relish multiplicity!

Yet it is this very notion of multiplicity that Deleuze and Guattari seek to emphasise, and, as we are not hobbits, we should give them a hearing. Rhizomes are seen as networks and 'any point of a rhizome can be connected to anything other, and must be' (p. 7). The implication here is of connectivity and flexibility. Rhizomes have no centre and no necessary shape: 'A multiplicity has neither subject nor object, only determinations, magnitudes, and dimensions that cannot increase in number without the multiplicity changing in nature' (p. 8). They are flexible to the extent that they can be remade: 'A rhizome may be broken, shattered at a given spot, but it will start up again on one of its old lines, or on new lines' (p. 9). It is therefore 'not amenable to any structural or generative model' (p. 12). There is no predetermined pattern for a rhizome, which limits its possibility for developing and making connections.

This lack of any central co-ordination to a rhizome is crucial to its definition: it is what makes the difference between roots and rhizomes:

> To these centred systems, the authors contrast acentred systems, finite networks of automata in which communication runs from any neighbour to any other, the stems or channel do not pre-exist, and all individuals are interchangeable, defined only by their *state* at a given moment – such that the local operations are co-ordinated and the final, global result synchronised without a central agency (p. 17).

They go on:

> unlike trees or their roots, the rhizome connects any point to any other, and its traits are not necessarily linked to traits of the same nature … It has neither beginning nor end, but always a middle from which it grows and which it overspills (p. 21).

There is then a fluidity and unpredictability about the rhizome in contrast to the tree root, which is seen as hierarchical and patterned. Deleuze and Guattari suggest that 'The rhizome is an acentred, non-hierarchical, non-signifying system without a General and without an organising memory or central automaton, defined solely by a circulation of states' (p. 21). The difference is between loyalty and acceptance on the one hand, and an equality of purpose on the other, or as they state it, 'The tree is filiation, but the rhizome is alliance' (p. 25).

The tree signifies a starting point, the acorn or seed, and an end, when the tree dies. However, the rhizome is not about starting or finishing but offers 'another way of travelling and moving: proceeding from the middle, through the middle, coming and going rather than starting and finishing' (p. 25). They seek to 'do away with foundations, nullify endings and beginnings' (p. 25) favouring a continuing story that goes on and on, changing as it does so, but with no predetermined form and with nothing but contingency to drive the narrative.

This dichotomy can be seen as a critique of structuralism and a statement of the possibilities offered by post-structuralism (Boyne, 2000). The tree root is the metaphor for the rigidities of structural discourse. Indeed in their discussion Deleuze and Guattari often return to the linguistics of Noam Chomsky. This is doubtless because of Chomsky's frequent use of the tree metaphor to develop his conception of the construction of grammar, but also because his view is decidedly structuralist. However, it is not my purpose here to defend a structuralist position. What I wish to do is to dwell on the root/rhizome dichotomy in terms of what it has to say about difference and the extraordinary. In doing so, I wish to put up a defence for the notion of the root, as a significant metaphor for a structure, and also for the notion of rootedness in opposition to extraordinariness. What I wish to do is to show the need for roots.

My starting point is to suggest that Deleuze and Guattari are well aware of the caricature they are presenting. They state that the single root metaphor is an oversimplification and point to the fact that tap roots are multiple and diverse. This does not stop them from rejecting the root metaphor in its entirety. Yet we need to make the obvious point, that despite the emphasis placed on hierarchy and rigidity in their model, each tree is different, with its own distinct roots that service it. Even though trees have common characteristics and can be readily identified as oak or beech, they are still separate and can be appreciated for their individuality. As Thomas Pakenham (1996) demonstrates, there is no shortage of remarkable trees, with a history and a personality. What this suggests is that even if we are all of the same species, we are still subjects who can be differentiated. We do not therefore need to remove our roots to demonstrate our differences; we do not need to deny or denigrate what is common to the species in order to be unique.

But, of course, we can only understand what Deleuze and Guattari are proposing because we are well aware of what they are denigrating: what is 'post' can only be appreciated by a thorough understanding of what was before. And so *A Thousand Plateaus* only makes sense if we know what they are arguing against, if we are fully conversant with the root metaphor and take it as being ordinary and normal. The rhizome is only a radical departure if we are embedded in what it seeks to cut us away from. This, of course, is a standard critique of post-structuralism, that it resides in and comes out of what it seeks to oppose, and without that paternity it has no means of justifying its own status (Benton and Craib, 2001). So, the rhizome metaphor needs the root – the rational, ordered discourse, with a start, middle and an end – in order for it to be understood. Like all post-structural concepts, the rhizome depends on the root: it needs to be rooted into something solid for it to survive.

This latter point brings out the importance of the root idea. The root is grounded in something that is solid and permanent, which supports it physically, just as the root gives physical support to the plant. Plants, of course, can be pulled up, transplanted, re-potted. The root needs space and may need to be moved if the space is insufficient. Yet it cannot be in continual motion. It needs to be put back into something solid. Roots need to be settled.

But, of course, it is through roots that the plant is settled. Roots are not there for their own sake; they are parts of a plant with a particular purpose. One feels that Deleuze and Guattari are proposing the rhizome as an end in itself, that if we achieve a rhizome book, or if these networks do exist, then we have achieved something. Is it not a problem of post-structuralism that it favours process over outcome and, in doing so, ignores the instrumental nature of processes? Processes are not ends in themselves, but rather are necessary to achieve an outcome. The process may of itself be cathartic, but catharsis is itself an outcome beyond or additional to the process. Merely because something could not be achieved without the process should not lead us to favour process over outcome. We should judge processes by their outcomes rather than seeing difference as purpose enough.

Networks, however complex, are means to an end. What is important is *what* they connect up and not just that they do. So the very fact that the root is grounded means it can sustain the plant by providing nutrients. The root, indeed, is the main supporting element for the plant. But it is not the reason for the plant. The root is as it is because it has a purpose to fulfil, and it *can* fulfil this purpose. It is the plant's response to the demand for life. The root is what seeks out the nutrients needed for life. It is therefore active and purposeful, but it is not an entity in itself.

In this way, the root is the main conduit to the external world. It is the only point of direct and regular contact with the world outside. The criticism of Deleuze and Guattari is that this is the *only* point of contact, unlike the rhizome with its multiple entry points. Yet this does not diminish its significance. Indeed it rather heightens it. Whether a plant has one or many points of contact it still needs the nutrients, and the fact that the root is singular does not affect this. Of course, the criticism is really that if there is a sole point, this can lead to dominance and dependency. Yet any plant, by being fixed, is so dependent. Implicit in the model of the rhizome is a sense of non-dependency and of contingency. Having a multiplicity of entry points and lines of flow, a rhizome is able to make and remake connections. It can make alliances, which may be temporary and may change as circumstances change. Yet the relation between a plant and its source of nutrients is not a contingent one, but is rather a necessary condition. Quite simply, the plant must be grounded and remain so to survive. The relation is indeed filial in this regard.

The same applies to how we dwell in communities. We can change where we live, but we still need to dwell and to do this in a manner that retains the integrity of our household, nurtures those who depend upon us, and respects others around us. We do not live through contingency, but through regularity, consistency and that which is known and familiar. We ourselves are dependent on supportive structures, which are physical, social and emotional. But these help us to thrive, rather than to dominate us.

We are rooted in what is around us. We have settled into it. We have become, as it were, anchored into the soil, into that place which is ours, with its particular rhythms and the sense that flows from it. This rootedness is found in institutions such as family, community, class and nation, as well as in particular places that we have come to love.

Being part of the whole does mean a dependency, but an exclusive one. The root needs the trunk, the leaf and the branch, just as they need it. The plant is not solely dependent on the root. The plant is contiguous: the root is but a part of the plant. There is literally an organic unity here, and this is seriously underplayed by Deleuze and Guattari in their comparison of root and rhizome. And this unity is an ordered one, in which the sense and pattern is quite clear.

The main criticism of the arborescent system offered by Deleuze and Guattari is that it is hierarchical and has some 'General' or 'organising memory' to co-ordinate actions within the system. The implication of this is that it is oppressive or representative of a reactionary form of structure as opposed to the acentred multiplicity of the rhizome. This criticism seems to amount to the accusation that root systems are ordered, whilst rhizomes are not.

But what would it be like to be part of a decentred network? If we had nothing to focus on apart from contingency and flux, would we be able to live in a measured manner and understand our place in it? I would suggest that this makes no sense at all: even if the world were really as Deleuze and Guattari suggested, we would still need the illusion of a centred world, for how else could we operate as rational beings? If we knew there was no purpose, that there was the possibility of change at any moment, how could we plan? If there was no sense of the long-term, would we feel able to get out of bed?

The idea of the root is the very opposite of this vision of flux. It offers a sense of certainty and integrity. The root is a symbol of continuity, of a located fulfilment of our plans: by staying still and remaining constant we grow. We know what it is that nurtures us and we reach down into it, and we know that if we do so – if we lean on our root – we have the chance of becoming strong, of developing and passing a part of us on to the future.

What this means is that *we* are the 'General'; the organising memory is ours. And this is suggestive not of oppression, but of liberation, of an ordered sense of freedom where protective structures allow us to plan. Accordingly, I reject that we can ignore our rootedness, or pretend that we are otherwise than rooted beings. We have a root and without it we are undernourished, we wither to nothing. And, what is more, we know it, for how else are we understood?; how else do we connect up with other beings?; how else are we nourished?

Ruts

Roots suggest stasis, that we are fixed to a point. Indeed, this is the very great benefit of being rooted. Yet we do need to move. We need to go from place to place, as part of our daily business. We are active creatures and we seek to explore.

Yet the majority of us put strict limits to our explorations. We may wish to travel far and wide, and on occasion we may be able to do so. But for most of us and most of the time, we are tied to persons and to things. These ties cannot be neglected because of the dependencies they bring. It is true that we have sought out many of these persons and things: they are our family, our friends and our belongings. Yet they do then bind us to a place.

So when we do travel, we may not travel far. And because of these dependencies we tend to travel in known ways and along known routes. We go from particular place to particular place, to do regular and known things. We go to and from work, school, or shops, to do those things that are common to us. What I would suggest is that we travel in particular *ruts*. These are tracks made deep and permanent by the regularity of our journeys. They are the routes we habitually take. Moreover, they are the routes we share with others. They are the known paths of a shared life.

These ruts are not a multiplicity. Truly there may be very many ruts, and they do indeed connect up as an apparent network. But this is not a network of pregnant possibilities. Rather what we see is a number of set ways, that over time have become carved into the landscape. Ruts form through use: we travel through them, but also they are formed because of our need to travel, and to travel *that way*. It is because we have to go to a particular place or thing, and there is no other way. This may not be because there are no other possible paths. Most likely it is because the rut is already there: we choose that way because it is already formed and therefore we know it will lead to where we wish to go. As with roots, we do not form ruts – we do not make a pathway – because we like them or feel they have an intrinsic worth. Ruts form because of where we want to go. Thus the possibility of a new route does exist, but only for the time it takes for a particular path to form, and as that path forms into a rut, and becomes deeper and deeper, the possibility of an alternative lessens. This does not, and need not, prevent anyone from striking out on a new path. Yet they do so knowing there is an available alternative, where the risks are known and the terrain is settled. And we cannot help thinking why should we strike out anew, when there is a rut we know that will take us where we wish to go?

Ruts widen with use and expand to become something that can be shared: a road begins as a rut. The more it is used the more it will itself develop and change, even as – and because – it still takes us where we wish to go. The utility of the tool that takes us to where we want to be, becomes more evident the wider it becomes. And there is safety in numbers as we travel.

But, of course, there are problems with ruts. We can, proverbially speaking, get stuck in a rut. The implication is that we no longer seek out new possibilities and become stale. What we need, by implication, is something new, to take a different route to get to where we want to be. This is doubtless a risk, and ruts certainly do limit our possibilities. But for many of us being stuck is precisely what we want. We want to be in a known place, going along a known path in a safe direction. We have to, for otherwise we would lose our way and have no means of knowing where we were going. More so, there is a virtue in being stuck in a rut, to be sure that our wheels go along a particular groove. This allows us to concentrate on other things that are more significant: we can look forward to getting there rather than having to steer an unsteady path. When we are in a hostile terrain we look out for paths already made. This gives us a certainty and a confidence in our journey, of something well-trodden.

As I have suggested above, what would we do if faced with limitless possibilities; with a multiplicity of avenues stretching out into the distance, their twists and turns offering us nothing but uncertainty? We need to know where we are going and what we will find when we get there. We do not need nor want an infinite number of routes in order to get there. We may choose adventure, but this usually leads us to increase the insurance we take out as a precaution. It also causes us, and those who care for us, to worry. In contrast, the rut is safe, it is secure, and it is known.

We live in one place at a time, and often we stay in that one place for a long period of time. We get to know that place and what surrounds it. We come used to certain routes to and from our place and other places where we need to go. We do not want surprises when we go from place to place: we want to actually get there, because what matters is what we do whilst we are there. In consequence, we tend to wear ruts into the ground; to make certain and true paths to and from home.

Accommodation

Our ordinary world consists of roots and ruts. We should see this in both literal and metaphorical terms. We are located, or stuck in the ground: we do go along particular routes and choose these again and again. We are also beings who seek certainty and who look to habit and the regularities of the past in order to anticipate the future. Contingency is all right in theory, *but only in theory*: it has no place in the ordinary world where actions have consequences and roles are fixed. We live in a world of rules, not all of which can be broken.

By emphasising the roots we have and the ruts we travel along I aim to show just how much we have in common. We live in an ordinary world in which we share much, and in which that sharing is an ordinary part. We do not seek to differentiate ourselves from all others, but rather look to ways in which we can draw together. Of course, we close the door at night to keep out the unwanted, but then so does everybody else. The concept of the amorphous rhizome is inimical to the ordered world we create, where we have everything in its place so we can take it for granted. Our dwelling is not a transitory world, where we are in flux, but one of stability, where we can nurture our children and care for our partners. Or at least this is what we hope it to be.

Our dwelling serves as a respite to change. We do not change place often and thus our roots grow deep and strong. We do not need to move often and we do not wish to. But when we do move, we do so along only a very few routes, which are known to us, which are safe. We live, I wish to argue, in an ordinary, usual and common place.

What I believe this discussion of roots and ruts centres on is the idea of *accommodation*, of being somewhere in particular. But it is also where we find an acceptance because we have been taken in by this place. This sense of accommodation comes when we have taken root, where the wheel finds the rut, where we share without complaint because we are being accommodated by those we care for and who care for us. Through this accommodation we learn to accept what is around us and come to terms, not only with what it offers, but also with its limits: by accepting the place where we are we are able to see it properly and for what it is. We are now able to see both the normality of this ordinary place, but also its complexity, beauty and splendour. This notion of accommodation, which, of course, has a double meaning, can be seen as the antidote to the hubris of post-structuralist thought with its focus on transformation and transgression (King, 2004b). But it is also much more, in that it offers a positive statement of how we live, not by change but by staying put.

This situation is shown in Peter Hammond's film *Spring and Port Wine* (1969), a gentle comedy set in a Northern English industrial town in the late 1960s. The story centres on a family of middle-aged parents and four grown-up children. The father (James Mason) is the centre of the home as the source of authority, and a figure to be both mocked and feared for his rigour and sense of certainty. This certainty oppresses the children, making them feel diminished and unable to assert themselves as they would like. On a Friday they hand over their earnings and the father checks his wife's housekeeping accounts.

Yet we sense that each person is still centred within the home. Each has a role, a place or purpose which he or she fulfils, like spokes around the hub of the father. They fit within a routine that establishes and ultimately satisfies them (as we see when none of the children leave home despite their apparent desire to do so). They may want to leave, but they are not aware of what it is that they want to put behind them. In one scene, the two boys pack their belongings, put on their coats and prepare to leave. But when they are asked where they intend to go it becomes clear that this had not even occurred to them. The two daughters also make plans to leave. The youngest daughter, in dispute with her father over a kipper she refuses to eat (her queasiness is due to her being pregnant), goes to stay with a workmate, but she soon returns disapproving of that family's habits. Being away for only a short time makes her realise what ties her to her own home. The elder daughter is planning to marry and goes to look at a house with her fiancé. But even this new potential home is shown as anonymous and uninviting with the furniture covered with sheets. It too lacks the feeling of specificity, the peculiarity that makes home.

What becomes apparent as the story develops is that regularity is the source of both comfort and resentment. The father sees home as a place of love, yet he offers no open show of affection to his children. For him, love is about instruction, discipline and control, over both himself and others. So he is stern and unyielding to his children, as in the case of the kipper he orders to be put in front of his daughter until she eats it: he feels she should be grateful for what her mother has cooked and that she should learn not to waste good food. He describes raising children as a battle between them and their parents, where the aim is to control the desires and wishes of the young that would lead them into trouble. The mother sits in between, listening sympathetically to both sides, defending the children when she can, as she sees this is the best way for the family to survive and thrive.

But when all this starts to fall apart – the children threatening to leave or actually leaving, and being helped by the mother who has raised the money by pawning the father's new coat – the father finally tells his wife why he takes this attitude to family life. His attitude to raising children is in reaction to his own childhood, where his dissolute parents struggled to pay the rent and maintain stability. He is determined to be the very opposite of his feckless parents. This, of course, is the great cliché at the centre of the film – *that we act against how our parents behaved*. His children react against his apparent rigidity: they smoke in the house; his daughter falls pregnant; they wish to leave *precisely because* he sets such great store on being together. Yet he has become the martinet to his children *because* of his parent's dissolute behaviour.

We can see that the film is about the resolution of this switching against parental force: it is about finding the medium between extremes of control and licence and demonstrates that parenthood is about finding compromise between sympathetic intelligences, where we accommodate the differences because of what else does and can bind us. The youngest daughter, as we have seen, readily returns to the family home when she sees how others live: as she says, 'They are not like us, are they, Mum?' The film is concerned with what is thick between us and how we can establish the space without breaching that connection. The thickness of blood can clog and coagulate and prevent us from carrying out our own movements. And so we might yearn for something that is thinner and clearer in which to swim. But we still need this medium: thick or thin we cannot survive without it.

What this shows is the need for accommodation, for the manner in which we can and do make allowances for those we know and love, and we do so because of what we share. This is a thoroughly illogical and circular process, in which the accommodation becomes dependent on love, and love dependent on the accommodation. So we see that the mother is desperate to show she can manage the housekeeping, and thus the 'window cleaner' becomes a balancing item if she cannot account for where the money has gone, and she borrows from the children to 'keep it right'. Yet we find at the end of the film that her husband has known all along that she had been 'fiddling' the accounts (not of course for her benefit, but out of financial incompetence and fear). He has pretended to accept her excuses and reasoning to save her face and to keep things as they are. He is therefore, despite his rigour and principles, capable of accommodation, and so the ending of the film, where he bends and accepts his children as they are, is not really such a step for him. His accommodation is rather to lose his fear, and to be able to express those doubts and anxieties that have driven him for so long; what he was finally made to admit was that his principles did not rest on certainty but on fear.

Within families and relationships we pretend and we hope that we are not found out. But much of this pretence can only be maintained by the active complicity of those we are trying to deceive. They let us get away with it to preserve the peace and to maintain the dignity of the deceiver. Or they let us carry on as the martinet because this is what they wish to push against. It is a form of paradoxical but mutual compatibility, sometimes based on fear of the consequences, perhaps of change or of losing someone, in the knowledge of the irreversibility of time and indeed of knowledge, and because we love that person and do not wish to bruise their dignity or insult them. We respect them so much we want them to continue in a manner that does not diminish them, even if this might cause us stress and worry.

Love, then, is about acceptance and accommodation, not just in the sense of being together, but of not rubbing against each other. We live together and in doing so we give something up to achieve a common end. It is a mutual acquiescence, which at its best becomes a single thread. Perhaps more commonly, we believe we have enough to keep us together and wish to preserve it for each other. The crucial element in all this, as shown in *Spring and Port Wine*, is not just where we live, but how and with whom.

This film also demonstrates that roots and ruts do not prevent change; they merely limit it, channel it and direct it along particular ways. They provide us with a set of limits within which we must seek some form of accommodation. There will be conflicts, and occasions when we rub up against the boundaries, or indeed seek to push them over. Sometimes we will indeed succeed – the wheel will leave the rut, we might be uprooted, or a cutting taken and replanted to form something new. Yet where this occurs, what we soon see is the formation of new ruts and the need for new roots, for without these we cannot be accommodated.

We can accept certain people and the situations they create because we know them and love them. We can be hurt by them, and feel misunderstood and unappreciated. These are feelings that we might have as children, in the face of apparently unsympathetic parents, but also as parents ourselves, when we feel our children do not appreciate the sacrifices we have made and the sheer time and effort devoted to nurturing and protecting them. Yet this does not diminish the binding between parent and child: the apparent lack of appreciation does not prevent further sacrifice without question, nor does the apparent lack of sympathy prevent us from constantly turning to our parents for succour. What we have here is a practical obligation, which may be unspoken but is also unquestioned. We are tied in and we know we are, and we wish to be, and all this operates at the level of ordinary experience. Perhaps this is the way we should interpret the Hegelian notion of freedom, whereby we are only deemed free where there are institutions and boundaries that bind us, and which we are able to identify with (Hegel, 1991). It should perhaps then not be a surprise that Georg Hegel built up his philosophy of right from the bottom and from those institutions like the family and private property that are centred round personal obligation.

We are not decentred beings, and we do not relish flux and contingency. We need a place to return to, even as we find it restricting and that those whom we live with are unsympathetic or do not appreciate us. As Hegel suggests, we do need to restrict ourselves in order to be free, for without these restrictions we can do nothing. We need to accept others, just as they accept us, and this mutual accommodation gives us something that is both definite and certain. We are located firmly and definitively in concrete relations.

This situation, I want to suggest, is something to be relished rather than condemned. We can all point to examples where there is no acceptance, be they cases of separation, domestic violence or child abuse. We know that these things occur and that their consequences are deeply and truly serious. Yet I want to suggest that one of the reasons we find these issues so troubling is precisely because of the manner in which it diminishes the sense of obligation upon which our private lives rest. I would argue that we see domestic violence as wrong for three reasons: first, that it physically damages another human being who is normally less powerful than the perpetrator and unable to defend herself; second, it is an infringement of the autonomy of an individual; and third, it breaches what we see the dwelling as providing. This is why we make a distinction between domestic violence and violence *per se*. We are disturbed because it questions the stability and wholeness of the ordinary domestic environment. Moreover, it disturbs us because it is the *exception*, it breaches the normal expectation of what private dwelling ought to be (King, 2004b). As exceptions, cases of domestic violence should not call into question the need for secure roots and ruts, but they ought rather to point to their importance. These cases are where the protected intimacy that comes from dwelling breaks down. And so we should not downplay child abuse or domestic violence: these are evils that need tackling, and the reasons we need to tackle them are self-evident: we can see what is lacking by just looking around us, at what we have, what we share and whom we care for.

The problem therefore is to deal with the exceptions without losing sight of the fact that they are exceptions: to keep a sense of proportion and not to let our commitment to deal with cruelty and injustice cloud the day-to-day experiences we have of how life is most of the time for most of us. It is easy to criticise most housing researchers for commentating on ways of life that they do not experience themselves, but it is also pertinent. Most of us, most of the time, live stable lives, and it is this very stability that allows us to look around and question how others live. The result is that we concentrate on those who have no stability and who are not well-housed, and forget that these are exceptions, and, in doing so, ignore what is important in our own lives. What is called for then is a robust case for the default situation, of the ordinary manner in which we dwell. This may not be how we all live, and we need not go into too many specifics. Rather what we need to make a case for is the manner in which we use our dwelling to get away from the exceptional and become ordinary. This is what this book now seeks to do.

Chapter 2

The Unseen Frame

The window frames our view of things outside, but we do not see it. We only see what is past it. We know the frame is there, but we look beyond it to what we wish to observe. But as I look from my window I see nothing unusual, nothing that is not ordinary. I see other houses, cars, a road, grass, trees, the occasional cat, bird and human being, all acting normally, doing what these things and beings do. And all these things seem so accepting of their place: there is nothing unusual or strange in what they do, how they do it and why they do it. They provide no surprises and expect none from what is around them. *Everything is just so ordinary.*

Yet this is the very stuff of life: where not much happens and it happens as it invariably does. It is predictable, quotidian and complete in its complacency. But, it is nevertheless fascinating and a source of wonder. Philosophy, said Socrates, begins in wonder, and this is a wonder that things are *there*, that, to paraphrase Martin Heidegger, there is something rather than nothing. But it is not just that it is there, but that *what* is there, and there like *that*, is of such beauty, such complexity, and yet it appears to fit together as a whole. For the ordinary is a whole; it is both a unity and comprehensive (Rosen, 2002): it is all that there is and all of it is of a piece. And like the window through which we view this quotidian scene, our ordinary world frames our actions, and we do not notice it. Our ordinary world is an *unseen frame*, and this is the special function it has for us.

But not merely unseen, it is also *unsaid*. The ordinary world can be that which is unsaid, because it need not be articulated in order for it to work perfectly; and because saying it – articulating it – might well alter it in some way. Likewise, once we have caught sight of it, it cannot go back into hiding. The ordinary is like one of those visual puzzles in which at first we see nothing, no pattern or regularity. It is merely dots or splodges on a page. But once we see the pattern we cannot then 'unsee' it. We cannot 'unknow' the patterns once they are made clear to us. We now know that there is a pattern instead of a random series of dots. Likewise, when we see those things in our lives as distinct, as separate from their surroundings, we might struggle to 'relocate' them again. So when we begin to look at the particular tree, an individual bird, see the human shape as our neighbour who takes in our parcels, we start to notice the specialness of things around us.

One house, we would think, contains much the same features and amenities as any other: rooms used for the same purposes, filled with machines, tools, furniture; spaces are connected by doors, windows, stairs; walls are full of pipes and wires. One house is like any other, an amalgam of bricks, plaster, wood and metal. It is surrounded by similar amenities: roads and pavements to join the houses together and which hide more pipes and cables; green space, both private and public; traversing humans and assorted animals, both wild and tame, but all common. We live in quite ordinary places.

But whose housing is ordinary to *them*? Is not *our* housing and *our* environment special to *us*, and does it not have particular qualities of comfort, warmth and worth that are shared with no other (except perhaps the home of our parents). It is a unique environment, made so by it being ours. Yet despite this specialness, many houses do indeed look similar. They have the same functional and performative capabilities as any other. In no physical or aesthetic sense is this particular brick box unique.

What we have, then, is a dwelling that is quite ordinary, one that is not unique or fundamentally different, *but* which remains special, unique, to those who live there. And, crucially, this specialness derives from the fact that it has become and remains our normal environment. The uniqueness of *our* dwelling is because it is what is ours and no others'.

I believe that this paradox, if that is what it is, rests on the meaning of the word *ordinary* and how we would use it. It is the difference between, on the one hand, *mundane* and, on the other, *being in the midst of* (Cavell, 1988). These are apparently very different senses of the word 'ordinary', of what is commonplace. What this differing sense rests on is the manner in which we view others and ourselves, or stated otherwise, the difference between *the common* and *the particular*. What is particular to us is, of course, always there. We are in the middle of it as it forms our everyday environment, and it too is therefore, properly speaking, common. Yet because it is *ours* and the story is about *ourselves*, we do not see it as such. It would be demeaning to our sense of worth to see it as common. Rather we see it simply as *ours* and this is what makes it different, very different, from any other dwelling.

But what we do not do is spend our time looking at what is around us. We use what is around as a picture uses the frame, as a wheel uses the rut. Yet it is also what roots us to this spot, what grounds us, anchors us, supports us, nurtures us and does so because it is always there. It is that taken-for-granted and unseen frame through which we observe the world, even as we lean on it and let it support us.

So as I sit and start to look – at the things beyond *and* the thing that enframes – I am staggered by it all and I begin to wonder. I know I have seen all these things so many times before. I have walked through this place, over it, beside it, engaged with it and at other times ignored it, and yet here I am marvelling at its sheer scale and variety. And all while it goes on by, ignoring me and what is mine. So what I must do, if I am to engage fully and make some sense of what I know is so ordinary, is to impose some order, some rigour, on what is just there. I need to impose some sense on what it is I am seeing. And as I do so, the ordinary stops being unsaid: now that it is seen we also see the need to talk and exclaim about it.

But there is an immediate problem and it is quite fundamental. What puts a rigour, an order, on the ordinary is ourselves. It is our lives, our unique, but quite ordinary lives consisting of normal pleasures, that provide the rigour. It is the very fact that we see sense in what we are in the midst of. We are truly thrown in the world (Heidegger, 1962), and this throwness is itself part of the world. We cannot separate what we are from what we are in: the unity encloses us and the comprehensiveness extends to include our lives. This means that we take the risk of losing something if we try to step outside of what we are in. Yet how can we observe it properly unless we stop and peep around the frame? Would we lose too much, would we become, as Stanley Rosen (2002) suggests, too rigid and lack the necessary subtlety to define what is (or should be) within us and which we are (or should be) within? And is there not also a legitimate fear that we might not be able to return, at least not without many scars, without knowing too much about what we might call the wrong things?

My fear, and yet my hope, because of what I think can be achieved through a study of the ordinary, is that a life is only as full as it needs to be and yet remains as empty as we can bear it to be. By this I mean, that when we are busy 'just living' we have no conception of whether there is too much or too little in our lives – whether the glass is half full or half empty – or whether we are missing out on much. What we are in the midst of is just the result of being where we are. This suggests that we can only contemplate our lives when we are either too full or too empty, in crisis or stasis. It is only when we lose what we have, or when we are struck with wonder (and, of course, it might as easily be fear, for what we have lost or might lose, or because it does not seem to be enough, or that it is too much …), that we look around. So, to wonder is not always a boon, but can be a burden, and we might not be thanked for encouraging others to look around.

But, let me ask, have we any choice but to look around? What would we – we ordinary humans – be condemning ourselves to by confining our wonder (or should it be stupor) to our navels? What would it mean to be just too busy, too self-concerned not to think about the complexity, beauty and sheer scale of what is just out there in open view, if only we would look? And as long as we do not stay away too long, as long as we do not lose our place, we have much to gain in raising our gaze up from our toil for a while and pondering on what is around us. So we should perhaps take the risk, along with the proper precautions, of course, and start to look at just what it is that is so damn ordinary around us.

The Magnificence of the Ordinary

Much of what is said about the ordinary, and it is more than we might think, is really an attempt to define the concept. Be it J. L. Austin or Martin Heidegger, Edmund Husserl or Ludwig Wittgenstein, each of them is sure the ordinary is fundamental, yet they find it hard, perhaps too hard, to arrive with any precision at just what the ordinary is. Or at least they cannot achieve this without losing us along the way. As Stanley Rosen (2002) tells us, this is partly because thinkers have found it hard to consider the ordinary without depending on abstractions, oblique turns of phrase and an over-sophisticated analysis, all of which is just too extraordinary to arrive at the sense of what *we* mean (because we know what we mean by the ordinary when we say it, but we seldom find this is sociology and philosophy books). Yet, as soon as we turn to the dictionary to help us, we start to glimpse something of the difficulty these thinkers have faced. The vagueness of one concept is here fleshed out by other words which beg qualification, further thought and take us on a circular tour of some other basic terms we tend to bandy around willy-nilly. The ordinary, we are told, is that which is customary or usual; or it is familiar, everyday or unexceptional; or yet, again, it is uninteresting or commonplace. This is undoubtedly helpful, but it is not unlike defining a cow as having a head in front, a tail behind and a leg at each corner! There is a lack of precision in our definition of the ordinary: what do we mean by customary, or usual or commonplace? Might we not just as readily define these terms by the word 'ordinary'?

Of course, it is precisely this sort of problem that motivated both Austin in his studies of ordinary language, and Wittgenstein on his quest to return philosophy to its proper status as the poser of the right questions in the right manner. We can see their philosophical endeavours as specifically aimed at dealing with such problems of circularity. But, as Rosen argues, can we really claim success on their behalf? Does not Austin's sophistication and Wittgenstein's obliqueness detract from their attempts at a clear delineation of the ordinary?

What this indicates is that there is a need for caution with language and with definitions. On the one hand, there is a need to offer as precise a definition of the ordinary as possible, but, on the other hand, we should attempt to do this without overstating the nature of the ordinary: we need to avoid some great theoretical construct, for this will abstract *out of* the ordinary. But if, along with Austin, Wittgenstein and others, we think that the ordinary is important, we must give the notion its full measure. What we have then is a concept that we see as thoroughly important, as pervasive and all-encompassing, but which is so fragile that we might endanger it by an overzealous exploration. If we dress it up too much with theory and highfaluting ideas, there is a risk that it ceases to appear in any way ordinary any more.

But more generally, a word that has so many cognates will inevitably carry problems with it. If we can easily shift from using one word (ordinary) and start using another (commonplace or usual), then we lose precision and might drift into a general and perhaps even sentimental exercise. I want therefore to stick with ordinary. This may not lead to a definition, but it will allow for a broader conception of what it is that I am aiming towards.

Perhaps a more pressing problem for any discussion of the ordinary is not just to include the usual, commonplace, regular, familiar and the everyday, but to suggest it can also be complex. What I need to do then is to distinguish the mundane and uninteresting from what we are in the midst of. On the face of it, this might appear to pose a considerable difficulty. The ordinary *is* mundane and uninteresting, precisely because it is familiar, regular and usual. It is expected, consistent and without unwonted variety. It is therefore apparently the very epitome of the uninteresting. But, I would plead that this is just not so. I want to suggest, and I hope I can show, that there really is much about the ordinary to be interested in, and this is so even when it is commonplace, regular, usual and so on.

The issue, I would suggest, depends on just whom something is ordinary to, and in understanding this, we will see some of the difficulties in defining the ordinary: the ordinary, so to speak, is always *mine*. When we say that something is ordinary we need to follow this up by asking 'to whom?' My life, like that of most others, is ordinary in that it is unexceptional, normal, familiar *to me*. Things take me by surprise and sometimes (and in hindsight) this is welcomed. But more generally, I have my routines and patterns, which may involve complex tasks, judgement and skill – marking, teaching, writing, supervising PhD students – but these are not particularly remarkable for me and for others with the same training, experience, skills, etc. This may well appear to be uninteresting and dull to a deep-sea fisherman, professional footballer, train driver or surgeon. But others – postgraduate students, less experienced university academics, etc – might aspire for this above all else. Indeed, many years ago these tasks would have seen little more than aspirations to me: they were things I hoped one day to be able to do, but which were out of reach. For yet others these skills might seem to be entirely beyond them. They are quite exceptional *from their perspective*. But then I would feel the same about a concert pianist, novelist, composer or cricketer. These individuals have skills that I cannot attain and it would be quite *extraordinary* for me suddenly to grasp any or all of them. Seeing a concert pianist such as Mitsuko Uchida play Schubert or a cricketer such as Rahul Dravid stroking a cover drive is inspiring; it is something I would dearly love to emulate (and in my dreams I imagine I have done), but I know that it is pointless to aspire to these things. I might consider that these skills are far superior to any I can lay claim to, but at least I can take solace in the listening and watching.

The point here is that something is ordinary *from my point of view*. This something might not be ordinary to others, and so we can say that the sense of the ordinary is not an intrinsic quality, but a relative one. It depends on the environment we are in, the skills we are able to master, our interests and other qualities, experiences and events. The actual causal relations and combinations are perhaps beyond picturing, but fortunately such a lack need not detract us from our course: we do not need to know all the ingredients to enjoy the meal.

What I suggest we need to do is to distinguish the idea of the ordinary as a criticism – as uninteresting – from that of the milieu in which we sit. I do not wish to criticise anyone for how they live or for what they see as usual and commonplace. What fascinates me is how this becomes and stays commonplace and familiar. I want to delve into that which we are all immersed in. This will be different according to who we are, where we are and what we have and do. And much of it will be similar – after all we do not see most of our fellow human beings as eye-poppingly strange or different from us. The ordinary is a relative sense that depends on our interests and capabilities. I want to maintain the idea, which I hope is not a fiction, that we all have our own ordinary, which covers all levels of experience, which interacts with others but is still *ours*. This, I believe, is important as it connects us but does not conjoin us: we are linked to others – most of us are commonplace enough to our fellow human beings – but we are still distinct and separate. If we were too distinct we would cease to be ordinary for others, and if too many of us were distinct then we could have no sense of an ordinary ourselves. There is to be no criticism involved in this: it is neither right nor wrong to be ordinary, any more than it is to be different. On one level then to say that we are ordinary is to say very little. It is trivial, in that we are just describing that which we are in the midst of.

But, then, because it is all we have, it is quite fundamental. We can see this by returning to Socrates' belief that philosophy begins with wonder. The ordinary is, so to speak, special because of its sheer extent. It matters because of what it lets us do.

Let me try to explain what I mean by this. It is the very *ordinariness* of the observed objects that makes them so distinctive. It is the stability of the perception of these objects that make them special, for it is within this perception that I can sit and be. It is through this perception of the ordinary – which acts as the frame for all I am and do – that I can sit and write, argue, breathe, doze and whatever else I choose to do. All of this is possible because of my knowledge, gained through my perception of things and of the stability of the ordinary, subliminal as it may be. It is that things *hold* which means we can act with some surety and security. It is the ordinariness of things that allows us to act. And a key part of this is the very ordinariness of ourselves, that we perceive ourselves as stable entities. This, of course, need not mean we are static, nor do I wish to engage in a debate here with post-structuralists or others about the nature of the subject and the self. What I mean is that our sense of the ordinary rests on the fact that, *at any one time*, we see ourselves and our situation as normal, familiar and usual.

So the ordinary need not be unchanging. Nor does it have to be mundane. We can see this by reflecting on the complexity and beauty of a flower bud. This is a thing as beautiful as it is common. Each individual flower bud is wonderfully constructed, layer on layer, showing a purity of colour and of form. Yet there are literally millions of them, too many to count, and so we can become quite blasé and ignore them. They appear regularly each spring and just as quickly disappear: they are little more than the background to our lives. But if we were to pick a bud off the branch and turn it this way and that and observe it, we might be captivated by the magnificence of the thought that something so common could also be so lovely. We wonder how it is formed, how it develops and then why it decays: how on earth has something so complex and beautiful developed and in that particular manner? And even a knowledge of the science of plants does nothing but expand the sense of wonder, as we realise it is even more complex than we had thought.

The ordinary is, then, complex, beautiful and constantly changing. It may have a regularity or a surface effect which gives us the appearance of unchangeability: the grass is green, the sky is blue and the clouds continue to pass over us. We understand the ordinary, largely perhaps because it has happened, and continues to happen. It is tactile to our perceiving of it. What makes it seem unremarkable is its closeness to us, that we live with it. As Stanley Cavell (1988) suggests, the ordinary is that we are in the midst of. We are in it, can walk over it, touch it, smell it and see it. It is the world that we are thrown into. We can choose to notice it, and then we may be struck with awe and inspired by what is around us. But to do this we need to stop in our journey and look. At this point the background shifts forward and the distance between our concerns and the world shrinks, so that we are surrounded by the magnificent ordinariness of nature. And at this point it becomes extraordinary because we are no longer, properly speaking, amidst it. It rather stands out in all its qualities, and we see it as apart from us.

So to return to the original doubt as to whether there a danger in stressing ordinariness, we should ensure that we separate out the ordinary from the unremarkable. The complexity and diversity of the ordinary world around us is in fact remarkable: it is very much a source of wonder. As in all things, choice of words is crucial. To state that something is uninteresting or unremarkable is to denigrate it. It implies we have little to comment about it, that we need not and perhaps even should not bother. And it is easy to link the ordinary with the uninteresting and the unremarkable. This is because it is all around us. We are in the midst of it, and we take it for granted. But what this tells us is that it is fundamentally important. It is the very stuff of our lives: it is what allows us to do and be. It is the bedrock, the foundation, upon which we rest. Without this sense of security we get from the ordinary world around us – of a place that is absolutely known and felt to be known – what have we?

One way of trying to understand this is to look at Ludwig Wittgenstein's conception of ordinary language (Wittgenstein, 1953). Speaking a language is something that nearly all of are capable of doing. We have learnt to do this through imitation and instruction and it is something that we continue to do, almost minute by minute. But we can do so without knowing, or needing to know, how we are so able. We are able to speak, be understood and understand others without knowing how this capability operates. We do not need to know the rules of grammar and the structure of language in order for us to be masters of it. More generally, we can operate in the world without needing to know why we do not float off into space or how we can see objects in front of us.

What the ordinary refers to, then, is what we are currently within, without an expectation of change (although, of course, we move from one 'ordinary' to another – we can leave an environment but we can only leave the ordinary on a temporary basis). The ordinary is the *normal state of affairs*. This can and does vary according to the case: what is ordinary to the senior executive of a global corporation like Shell or Coca-Cola is not the same for an elderly widow living in a sheltered flat. What is normal can involve little more than 'these four walls', the company of a cat and the small talk of neighbours, or it can relate to regular business class air travel, high-level meetings, luxury hotels, decision-making and the stress that attends all this. Yet even stress can be ordinary if that is what we are in the midst of: a stressful life can be the normal state. The apparently glamorous lives of some are still ordinary to them. But equally, we should not assume that the ordinary is dull, shallow or mundane to the person experiencing it, or for anyone looking in from the outside. Most of us live lives we like and we like them precisely because of the regularities embedded in them, and this applies to the thrill of risk taking as much as stroking the cat.

The ordinary is what belongs to our customary world. It is what we would expect from the world around us. This shows again that what is ordinary for us would not be for others – we cannot necessarily share in the ordinary sense of others, nor they in ours. We can sense this when we go to stay with friends and family. Like us, they live in a private dwelling, which will be decorated, furnished and fitted with modern appliances. Like us, they prepare and eat meals at regular times, have early and late rituals around the kitchen and bathroom and relax in front of the television and computer or listen to music. Like us, they behave as most families in the developed world behave. Yet in another sense, nothing is the same. Their taste in decoration and furnishing differs from ours, as do their tastes in food and choice of television viewing. Whilst much is the same as in our ordinary environment, everything appears to be different and we ourselves feel different because we are there. In this normal household that is not ours, nothing is ordinary even as most is recognisable. Things are out of synch with our sense of the ordinary and we look at the mundane activities of others as if they are somehow new and original. But, of course, and this is obvious by the way our hosts act, these activities are entirely normal and usual to them (of course, insofar as they are able to act with visitors around).

It is in these situations that we might recognise that our sense of the ordinary is a specific thing in itself. Seeing others doing similar things – things we recognise, like cooking, eating, socialising – but in a different manner, opens us up to the realisation that others may well look at *our* mundane ways as distinctive. Just as we remark on their funny ways, they too may see us as odd and mildly eccentric for doing something in our particular way. So interaction presents us with a sense of our particularity. And this is one of the reasons why I eschew the sociology of everyday life (Highmore, 2002), which seems to me to be an attempt to gloss over this distinctiveness we all have in our mundane pleasures. Instead of seeing us as diverse, there is a temptation to look for a 'big picture' explanation (class, power, culture) to show how we are all common because of what we all do on a daily basis. Yet it strikes me that it is only through isolation, by being separated atoms, that we would come to believe in our essential similarity with others, for in this situation we would have no recourse to actual experience to see that others can be and are different.

This interaction therefore tells us why we would seek out the ordinary, and why we relish the fact that we might be said to be ordinary ourselves. By placing ourselves slightly outside our sense of the normal, we come to realise what it is that we are in the midst of. It shows us that there is something – an environment, a relationship, a particular sense of ourselves – that is usual, expected and hence ordinary. By watching others in their ordinary world we begin to question what we have. By observing the many tiny differences that make us comment on the strangeness of people and how 'it takes all sorts', we come to understand that what we do is important and that it matters. But we can easily forget its import because we have no outside reference for much of the time.

Out of the Ordinary

Of course, in a very real sense, we cannot leave the ordinary. We are (nearly) always in the midst of the usual and the commonplace. This applies even when we visit others, for how else would we be able to compare how others live and act? We take our sense of the ordinary with us, and only rarely do we change. And when we do change all we can do is substitute one commonplace for another. For a time this new sense may seem to be distinctive and different, and we will notice the difference. Things seem odd, what we do is all too apparent and no longer habitual, and we remark on the difference. We might even relish the fact that things are distinctive. We may have changed things on purpose, by joining with a new partner, moving to a new area, or whatever. But soon, and without noticing, this new state becomes normal, established, accepted and taken for granted (even as we consider it a change well made). Once again we follow established patterns and routines, albeit different in some ways from what we did before. We again slip into the rut of particular ways of acting. We do things in a patterned way and we do not question, analyse or even really notice what we do. And we have done all this without being able to say when the slip back into the commonplace occurred, and perhaps not even realising we have sunk into it.

But do we still enjoy this new, but now established, pattern once it has become so commonplace? Can we enjoy something as much when it is tacit compared to when it was in the front of our consciousness? Must we be aware of something for it to be enjoyable? I would like to think that we need not be so aware, for surely the state of enjoyment does not depend on our being able to articulate that what we are doing is fun? We might be pleased that we have taken on a new challenge, but this is because we have achieved something we consider worthwhile. And for most, if not all, of us, it is the achievement and not the striving that matters. We take decisions because we feel we need to and make changes when they are necessary, but there is always an instrumental quality here. We are interested more in outcomes and not in processes. What is enjoyable therefore is what we have done and whether we can do it, not because of the possibility of a new adventure or transformation. Think in this regard of how a craftsman or skilled sportsman still gets enjoyment from their work or sport, even when they are grooved into actions that no longer require conscious thought. The batsman has ingrained particular actions so that they become almost instinctive. Yet when it all works, and he is playing the game well, there is nothing more pleasing. Also when we are reading an enjoyable book, are we not be more likely to see the contents of the book rather than the pastime of reading as enjoyable (even though they cannot be separated)? If we do not enjoy the contents we will stop and choose something else more interesting, entertaining or whatever. We do not normally just read anything – the phone book, a dictionary, cereal packets – just so we can say we are reading. Reading is enjoyable because we are reading something we like, and our newly changed but now commonplace environment is enjoyable because of what we do in it. We need not reflect on ourselves to be happy. Indeed perhaps we are happiest when we do not reflect and do not feel we have to.

When something is unusual or uncommon, when a book is not what we expect, or the batsman does something out of character, we describe it as *out of the ordinary*. It is something that is not within the bounds of what we consider normal. But what can it mean to be *out of the ordinary*? Is the thing *outside of* the ordinary, as in the sense of being beyond it? This would imply that it is in no sense ordinary, and may perhaps even be its opposite. It would be something beyond our normal expectations, in that it might be a much better experience than we expected, or that it might be so different – the batsman throwing off his normal caution and thrashing the bowling – that we could not possibly have predicted it beforehand.

But on the other hand, might *out of the ordinary* refer to something that comes *from* or literally *out of* the ordinary? It is a thing that has derived or developed from the ordinary itself. So we might find the book exceptional because it extends or further develops those elements in literature that we most enjoy: it is an exceptional *example* of what we most like. We consider the batsman's performance an exceptional one, but that is because we enjoy cricket and appreciate how it is and can be played.[1] Speaking simply, then, we need the ordinary to create anything that is extraordinary. And it follows that when something is out of the ordinary it does not leave the ordinary behind.

We can then be out of the ordinary. We can be in a state that is unusual to us. This may be troubling for us, or we may find it exciting: it can be worrying or exhilarating. But this sense of being 'out of' will either be temporary, as when we move house or suffer some trauma (and soon create new routines or impose our old ones on the new situation), or because we are one of that minority who consciously and actively seek out danger and the unpredictable. In the latter case, we actually seek to be put out, to face danger or inconvenience, and to be in situations that are not normal.

There is an immediate caveat we need to place here. When I say 'normal', what I mean is that state of affairs that most people conceive of as ordinary. This is behaviour that individuals would recognise as typical and dependent on routine and regularity, even where those routines and regularities are not the same as their own. We all have our own sense of the ordinary and, in recognising difference in others, see that a general sense of the ordinary pertains. Now this might appear to be merely an aside, or a convenient means for separating out us from eccentrics. It is, however, a major step in understanding what I take the ordinary to mean. Stanley Cavell (1980, 1988), in his work on Thoreau, sees him as an exemplar of the ordinary and the common. Yet Thoreau achieves this status not because he was a typical New England bourgeois and pillar of Episcopalianism, but because of his eccentricity that saw him build his cabin by the shore of Lake Walden and write about his life there. As Cavell suggests in the title of one his books, Thoreau was on a 'quest for the ordinary' (Cavell, 1988). Thoreau, as we know, was not considered ordinary, but was rather in search of it and attempting to create it for himself.[2] It is by exploring these exceptions, these actions that are *out of the ordinary* but come from it and, as we shall see, *accept it*, that will help us gain a fuller sense of what the ordinary is.

We can presume, if only because of the nature of human physiology, that explorers and soldiers in the field have routines and develop some sense of normality. They have their own habits that they have established and do without conscious thought. In addition, there are certain actions that they must undertake in order to survive in the hostile environment in which they find themselves. But they are still, we would think, in extraordinary situations, in positions where most

---

[1] As a corollary to this, someone who does not read literature or watch cricket would not really see what we are getting excited about.

[2] This raises a serious problem about how we can write about and present the ordinary. I return to this when I discuss Rosen's notion of the ordinary in the next chapter.

human beings would feel stressed and anxious for their safety. How then could they have any sense of the ordinary here, despite whatever routines and habits they might adhere to?

But might it not be the case that what is different between an explorer or soldier and our own sedentary lives is merely the time frame involved? Might it not be that what we could call *the cycle of routinisation* – that period that encapsulates the range of activities and reinforces them, thus creating and maintaining habit and a sense of ordinariness – is just shorter, so that the shift into the commonplace is less protracted. In this sense, we could see the readiness and indeed ability of explorers and soldiers to adapt as an ability to habituate more quickly? They are simply able to come to terms with their changed situation more readily than others. We might hypothesise that these people are able to *internalise* the ordinary rather than it being dependent on external cues and objects: that they carry the ordinary with them rather than needing to have it around them? *It is in them rather than they in it*, and, as a result, they are able to adapt much more easily. Put another way, they really have much less to change because they still have much of what is ordinary within them.

Of course, for all of us, our sense of the ordinary is made up of internal and external qualities. Therefore the speed and readiness with which we adapt to change is likely to be a question of degree, rather than being an either/or. We depend on the external to a greater or lesser extent, and the same applies to our internal sense of the ordinary. If this is so, it poses a further question: can and do we shift along a continuum between external and internal (which would imply that environmental factors dominate in the formation of the ordinary: explorers and soldiers adapt because they are faced with a particular situation), or do we tend to hold to an absolute position, a stable point? If we hold to a stable point, this in turn suggests that we would not necessarily adapt and that choosing to be placed in a particular situation is a matter of disposition. This would mean that the 'stable point' itself would be a part of our sense of the ordinary. This stable point would to a degree determine how prepared we were to experiment with the new, how dependent we were on home and family, or whether we yearned to be free of external encumbrances.

I have deliberately referred to a stable rather than a fixed point on this continuum between the external and internal. It seems to me that we alter our stable point with time, but that *at any one point in time* it appears to be fixed. We believe that our sense of the ordinary is as it should be, that it remains static. In short, we feel comfortable. We might not consider we have altered, but if challenged, however, might agree that we are now more home-centred, more staid and less adventurous than we once were (or, of course, the opposite).

But not only is our sense of the ordinary not fixed on one place on a continuum, it is also made up of a series of layers, so that apparently contradictory feelings can co-exist. So we yearn to be free, whilst still feeling a strong sense of external dependency. We might often crave little more than to be left alone in our favourite, most comfortable places. Yet, at the same time, we might like to know that we can be in ready contact with our family and colleagues. We do not wish to be isolated, just to pick and choose our points of contact and be amongst what we find most comforting. Or we might enjoy travelling to exotic places, making the minimum of preparation in advance. Yet we might only do this if we can go with someone we know well and can trust.

Having said this, there are indeed people who do genuinely appear to relish being out of the ordinary, for example someone such as Ellen MacArthur, the solo yachtswoman. She seems to crave the exhilaration *and* solitude of a speeding yacht in the open ocean. In this situation she cannot be comfortable (as most of us would see it) and there is nothing predictable, except that life is going to be hard. So where is Ellen MacArthur's sense of the ordinary? Now of course she is an exceptional person, doing things that most of us could not, and would not wish, to do, even as we admire her for achieving them. It may be then that she is one of the very few who has no need of a customary sense, of a sense of the ordinary.

So might not her craving for solitude, for challenge, for the openness of the oceans, be a form – albeit perhaps a perverse one – of custom? Might it not be an attempt at habitual behaviour and hence her apparent continual need to challenge herself? Might this be (and I am speculating here) because her sense of the ordinary needs to be freed from external clutter, to achieve a situation where she is able to rely on nothing and no one but herself? Her stable point is just at one particular extreme, whereas most of us would tend towards the other end of the spectrum. But she still maintains a sense of the ordinary, even when she is in the midst of a storm in the Southern Atlantic. She can do this because she has the skill and competence to deal with most things. Her sense of the ordinary might then derive from the internal, but it is well honed. It manifests itself as a series of skills and abilities (tenacity, stamina, determination) and does not depend on the obvious comforts of the environment: her ordinary is to fight the environment in which she chooses to reside.

More trivially might we view celebrity – being famous and revered for whatever reason – as still ordinary for those who are used to fame? It is ordinary, or becomes so, for them, but with the appearance of being out of the ordinary for the rest of us. It would not be particularly normal life for us, otherwise they would not be celebrated, but for them it has become something they are accustomed to (and therefore they feel they might not be able to live without it). We cannot share in their life, or at least we can do so only by gawping and by catching illicit glimpses, the process of which merely demonstrates the distance between us and them.

What we, or some of us, find difficult is to treat celebrity with indifference: we cannot ignore it. But the relationship between the ordinary and indifference is not a straightforward one. Who do we normally show indifference to? We might think that we are indifferent to those who are ordinary: we are indifferent to the mundane. Something we see as ordinary might not interest us, even to the extent of our not noticing it. But the ordinary is that which *we* are in the midst of: it is what surrounds us, including those people and things we live with. By this argument we would be most likely to be indifferent to those we share most with. They, after all, are the very opposite to the celebrities we may glimpse from a distance.

A moment's reflection, however, tells us that this is not the case. Those closest to us are generally the ones we love and care for most. Yet these are the ones who are part of and share our sense of the ordinary. This leads us back again to the question of what we mean by, or rather how we use the phrase, *out of the ordinary*. The people we live with are special to us, but often this is unarticulated. We do not continually dwell on the specialness and closeness of the relationship. Instead these relationships form the background – the ontological security – that allows us to carry out our daily tasks and seek to attain our own ends. We take these people for granted and this allows us to do the practical day-to-day things that are in the foreground of our lives. They are, then, there to maintain much of what makes our lives ordinary, and we help to fulfil that purpose for them. But if we were questioned about these people – and more pressingly, if we had to face up to their loss (King, 2004b) – we would readily admit to our strength of feeling: these people, whom we may take for granted, are the most important things in our lives. It is just that for much of the time we do not need to dwell on this. Indeed to do so would probably mean we were not functioning properly. But, at the same time, these are the people – *the only people* – we would live with. We see them as exceptional people, whom we wish to be with, and we are prepared to make sacrifices for them, to dwell on their sayings, thoughts and actions. And, whilst we do all this, we will happily take them for granted. Inasmuch as we love them, we also expect them to be there beside us: even as we see them as exceptional, their habits and actions are entirely predictable to us. Some things – or rather people – are inside our sense of the ordinary, but we see them as remarkable and are not indifferent to them. Indeed, it is because they are within our ordinary, that they share it with us, that we have this strength of feeling. Jung (1982) suggests that we attach the ideal attributes of the archetypes 'man' and 'woman' onto our partners: they become the very epitome of man or woman because we love them, and not vice versa. We need not accept the full implications of Jung's argument to realise the essential point, that closeness – the sharing of the ordinary – breeds a deeper sense of love and affection.

So we can be (at times) indifferent to those we are closest to, whilst taking a keen interest in others precisely because we have little in common with them. There is then a paradoxical quality in the ordinary that undermines it as mundane and uninteresting. There are some we love deeply and take for granted, others we do not know and show perhaps an over-keen interest in, whilst the majority just pass us by.

Of course, this general sense of indifference is a necessary factor in maintaining the ordinary sense of others. Whilst the life of a celebrity may have its pleasant side, it does offer a rather etiolated form of the ordinary: it lacks one quality of the ordinary that I would claim is of prime importance. The ordinary allows us to have the very opposite of celebrity, namely anonymity and indifference. Celebrities must purchase this commodity and thus be dependent on the regularity of its supply. This may work very effectively for them, but it is always something that must be imposed on them and on others: their privacy must be enforced. However, most of us can rely on mutual indifference. We do not actively dislike our fellow humans and, generally speaking, we wish them no ill will. Instead we are so full of our lives, so concerned with doing what we need to do for our loved ones and ourselves that we pass others by. If they fall we will help to pick them up; if they ask for directions we will do what we can; when they ask for change we might respond positively. But we do not seek to interfere in their lives and do not expect them to delve into ours. We want to be able to discount them from our calculations and hope that we too can be ignored. Without this we too would be like the celebrity, forever having to check their security, never able to travel alone and without advanced organisation. We might get used to this and it might become normal, but would we really want this?

A related question is to consider just how the ordinary relates to the manner in which we control ourselves: can we maintain a sense of the ordinary if we lack control? This refers back to the discussion on Ellen MacArthur and her adventurousness, and how there can be different senses of the ordinary. An individual like MacArthur may feel that she needs to be on an ocean-going yacht to be in control of her life. But most of us control our environment through mundanity, through not taking risks, and operating within clearly known boundaries. We maintain a sense of control by not pushing the boundaries, or stepping beyond them. And is this not precisely what the ordinary is: *a sense of operating within boundaries*? But, of course, the ordinary is precisely where we operate in a tacit sense, or through habit, and hence we are not conscious of these boundaries. We just live in and with what we have.

The ordinary has the wonderful quality of being both banal and fundamental. This is reminiscent of Martin Heidegger's obsessive concern for being, of that fundamental and fundamentally absurd question, 'Why is there something rather than nothing?' or 'Why is there being?' (Heidegger, 1962). One cannot persuade a practically inclined person of the point of this obscure question, of why it is fundamental and worthy of a lifetime's study. How can we persuade those who plead 'common sense' and 'of course, things exist and why not just accept this?' that Heidegger's obsession is anything other than silly? Accordingly we might ask, why bother with the ordinary when we can just do it, and when, indeed, we are just doing it right now.

But this is exactly the point: it is this very practicality that makes it special: the 'just doing it', but also the *only doing it and nothing else*. We can do nought else than live this way and it is precisely this, of course, that makes it so deserving of our wonder. The ordinary is nothing less that what we are, what we have, and what has us. We may feel that we are discussing rather little when we consider the ordinary, yet we are encapsulating the major part of our lives, be it professional or personal, practical or emotional. Ordinariness apparently does not inform us of much. It certainly does not distinguish us – it rather does the opposite. It tells us little of what we, as particular individuals, are and do.

It is this very anonymity, this lack of particularity, that is so special about the ordinary. It is that which allows us to live, in the sense of leading purposeful lives. The ordinary is the background into which we can hide and fit ourselves. We fit into the ordinary and, in doing so, we can seek to fulfil our ends, those things particular to us. But, of course, those ordinary things we do *are* those things most particular to us. It is both because we literally do them, and also because they so consume us and dominate us, that there is little else to us but this very ordinariness. It is these little details that consume us and make us what we are. They are at the very centre of our lives, comfortingly anonymous to others, but the very essence of us and ours.

We all have a sense of the ordinary, but for each of us it is experienced differently. The most universal expression of this distinctiveness is in the way we dwell. We live as private beings, albeit in similar ways. So we can point to a common activity, and therefore we might say it is ordinary: to dwell is commonplace, usual and creates many of the key regularities in our lives. Yet, whilst there is this commonality, we all do it distinctly, in our own way. Hence, as I have suggested above, we are never truly comfortable in someone else's private space, even though they, like us, have regular family meals, watch television and talk to each other. I would suggest that we extract this dichotomy and extend it to the ordinary. Accordingly, we can suggest that the ordinary is common *and* diverse: it is all around us *but*, at the same time, distinctive to us.

I am aware that this raises an objection. I am claiming that the ordinary is near *and* far, that it is specific *and* ubiquitous, and that it is everywhere *and* just here. Could I not therefore be criticised for attempting to invent a notion that acts as a coverall? Is not this discussion merely an attempt to develop a meta-theory as an alternative to other more developed sociological and phenomenological systems? By claiming so much for the ordinary, is there not a danger that it becomes an all-inclusive theory of the everyday? We can only resolve this if we are able finally to arrive at some conclusions as to what the ordinary is, and this is the task I undertake in the next chapter.

Chapter 3

Seeing the Ordinary

A Way of Seeing

I have stated that the ordinary consists of roots and ruts. The ordinary is about our being *in a place*, that this place is *known* to us, and that the manner in which we can move around it is *limited*: there are only certain ways we can and would choose to go. Yet that this is quite a wonderful state to be in. The talk of roots, ruts and limits should not be seen as reactionary or constraining, but rather as constitutive of the requirements for a life: roots and ruts make an ordinary life, and do this through proscription.

I have tried in the previous chapter to state what the ordinary does for us, and why we should relish this state. But what I have not done is to provide a definition of the ordinary, to state in a categorical sense just what it is. I have made a comparison between the idea of the mundane or the common, and that which we are in the midst of. However, it too became clear that the common *is* often that which we are in the midst of. This duality – very much an artificial one created for the sake of progressing the discussion – merely denotes different ways of seeing what is around us. What this difference amounts to is an *acceptance*: are we prepared to accept the condition we are in? In many ways what I wish to describe is not a theory, but a particular way of seeing what is already there around us. I do not want to develop a theory of the ordinary, for this would merely be an abstraction, but rather to consider some of the properties of ordinary experience. The import of these properties will allow us to make some statements about the manner in which we operate within our environment.

I wish to suggest that the ordinary is not so much a condition as a disposition, as a way of seeing. This disposition involves our doing nothing but just sitting and seeing the things around us in a particular manner. In so doing we validate and confirm the substantiality of the world and our place in it. As Stanley Cavell, one of the thinkers who has most developed this sense of the ordinary, has stated when referring to his observation of an object:

> I do not make the world that the thing gathers. I do not systemize the language in which the thing differs from all other things in the world. I testify to both, acknowledge my need for both (Cavell, quoted in Bernstein, 2003, p. 135).

We do not make our world, or its constituent parts; rather we testify to it through our actions, descriptions and sightings of things in the world.

However, what we cannot do, I believe, is to come up with a definition of the ordinary. This is partly because the tools that we have to define the ordinary – words, concepts, notation or whatever – are already a part of the ordinary. Thus the process of definition, the creation of a theory, would be to abstract from the ordinary and present us with only an etiolated sense of the concept. But there is a danger in taking this approach. In eschewing a programmatical approach to the description of the ordinary, I may leave the notion vague and insubstantial. In this sense, I am very aware of the problem that runs through Stanley Rosen's book *The Elusiveness of the Ordinary* (2002), where he charts the largely unsuccessful attempts of philosophers to grasp the ordinary. Philosophy, Rosen argues, tends to destroy the ordinary, by making it theoretical, abstract and etiolated. Even the deliberately unprogrammatical methods of J L Austin's ordinary language philosophy, and the descriptive therapy of Ludwig Wittgenstein are far too sophisticated and abstract to truly grasp the notion of the ordinary as common, everyday experience. The act of description, as it were, makes the ordinary special. As Rosen states: 'The ordinary is accessible, but it has the disconcerting feature of turning into the extraordinary even as we grasp it' (2002, p. 2). For him, the problem of analysing the ordinary is one of either turning it into a 'technical artefact' or remaining 'at the level of impressionistic and arbitrary anecdotes' (p. 8). Where it becomes too theoretical it falls prey to technical jargon, sophisticated abstractions which empty it of much of its common usage. On the other hand, if we rely on personal anecdote, we lose the possibility of generalisability, which may seem important for such a common condition as ordinary experience. Neither of these alternatives is sufficient in themselves, but if in doubt, or if we had the choice, I would prefer the anecdotal over the technical; at least this is *nearer* to ordinary experience. However, as we shall see, we do not have to rely on this extreme. We can make some more general statements on the ordinary, even if we do not seek to develop a theory.

Some Properties of the Ordinary

Rosen does not offer even a loose definition of the ordinary. But he does offer some key insights that take us forward. In particular, he seeks a grounding for the ordinary and to develop what he calls some of its properties. His starting and stopping point in considering the attributes of ordinary experience is *common sense*. Rosen defines this simply as our normal, standard judgement of what is good, right, for the best and usual. However, he argues that there is no structure to common sense, and therefore we can have no hard and fast definition of it. Common sense should be seen as the most rational or plausible thing to believe, given what we already know. Common sense is what we appeal to when considering whether an action or belief is ordinary.

But the ordinary cannot be all there is: there must be things which are not ordinary. As Rosen argues, we can appeal to common sense against eccentricity and extreme views, but if there was nothing but common sense there would nothing but orthodox views. Without *some* eccentricity and extremity there could be no change and we would have nothing but sameness. Clearly this is the situation that pertains. But neither does this limit the ordinary. Indeed, properly speaking, it is the ordinary that places limits on the transgressive. Rosen suggests that the properties of ordinary experience do not disappear when we shift to the extraordinary. According to him, 'ordinary experience continues as the context within which the extraordinary occurs' (p. 264). Rosen suggests we have ordinary experience as a 'common dimension to the discontinuous series of disruptions' (p. 264). The ordinary 'refers to the ongoing continuity of experience to which we appeal, often without noticing or calling attention to it' (p. 265). What this means is that the ordinary, so to speak, locates the extraordinary and gives it a regularity. So we can see the ordinary as a series of separated and distinct acts connected by some common thread. These commonalities can be seen as rules or laws which regulate experience, and because of their common application we can see them as usual and common. Hence Rosen can suggest, 'In general, we can say that, whenever there is a rule or a law that governs some portion of experience, the ordinary is that which follows from the correct application of the rule' (p. 265).

The essential point of Rosen's discussion here is that we can never lack the ordinary. It must always be there either as something we are currently in the midst of, or as something we can actively compare against other actions, beliefs and things. The ordinary is therefore the benchmark, or the default case, through which we assess the extraordinary. Therefore the extraordinary does come *out of* the ordinary in a literal sense.

This, of course, has important consequences for theory itself. If Rosen is correct (as I believe him to be) it is not possible, unlike what is argued by post-structuralists and deconstructionists alike, always to be extraordinary. It cannot be the case that all is difference. This would merely be a different sense of the ordinary. In order for there to be anything extraordinary, there must be something that is ordinary. Every extraordinary episode 'draws its distinctiveness from the contrast with the ongoing ordinary experience that serves as the backdrop for their [i.e., those who point to the extraordinary] pursuit of disruption' (p. 291).

Perhaps we need to go so far as to say that for something to be extraordinary, the majority of the world must be made up of ordinary things. There must be a rule or law which holds in most cases, in order to make another case an exception and therefore extraordinary. A life made up completely of the extraordinary, of pure difference, would be both unliveable and absurd. There would be no grounding and no limits which determined action. There cannot, therefore, only be difference: it has to be different *from* something else.

This does not mean, of course, that there can be no difference, merely that in order for things to be different they must be different from something, and more so, that the ordinary is the basis from which the extraordinary leaps:

> It is therefore no refutation of ordinary experience to call attention to variety, historical change, or hermeneutical difference … Nietzsche, Heidegger, and Derrida all step forward from the same everyday world in order to assert the priority of difference: in fact they never leave it, since they sit at their desks and employ pens, typewriters, or word-processors in order to compose testimonials to chaos, or, more modestly, to the infinite regression of interpretative standpoints (Rosen, 2002, p. 294).

But the ordinary rests not merely on common sense, but also on there being a human nature that we can rely on as a determinant of our actions. The ordinary depends on our senses, and our physical potential and limitations. So it follows that the ordinary rests on human physiology and there is no way around this. In consequence, any study of the ordinary must begin with these physical limitations: we cannot wish away what makes and keeps us human, and any theory or description that attempts this will be misleading.

From these premises Rosen points to three linked properties of ordinary experience, which he defines as regularity, unity, and comprehensiveness. He suggests that we can never experience the beginning of the ordinary (even though we may experience it coming to an end, if not exactly having ended). Ordinary experience is something that we are always within, and accordingly 'we cannot understand life by detaching ourselves from the flow' (p. 264). We are a part of what we are experiencing and therefore, along with all other beings and objects, we form a unity with it.

Like our dependence on the limits of physiology, experience depends 'upon the coherence of space-time and the regularity of the natural order' (p. 267). This though does not relate to intentional acts, which will vary. But they will only do this within the bounds of space-time and the regularity of the natural order:

> I am therefore always unified and orderly or regular in my natural existence, and this plays a large role in the unity and regularity of ordinary experience. But within ordinary experience there are many variations upon the intentional expression of cosmological or natural unity and regularity. These variations themselves follow general patterns of conduct, but the patterns can be violated. What makes life possible, and gives us the kind of experience we actually have, is that the violations are infrequent. Ordinary experience is what usually happens (p. 267).

So the fact that there is regularity enables us to vary from the norm. Variation rests on the normality of the ordinary.

Likewise, when we are making our lives – when we take decisions and act on our beliefs – we do so through a unity of experience, which we cannot dissociate or separate into smaller elements:

> it is impossible that human beings constitute the continuum of our ordinary experience out of simpler elements. It is this continuum that makes possible our discrimination of elements such as objects, actions, or events as distinct elements of experience. Drawing such distinctions does not dissolve the continuum but rather depends upon it (p. 267).

What I take this to mean is that we cannot decontextualise either ourselves or an object from the world and see us or it as divorced. We are always, strictly speaking, within the milieu of the ordinary and must take it as a whole. Rosen makes the point that we do not see experience as 'a process of the registration of stimuli upon sensory and cognitive machinery' (p. 272). Instead he concurs with Martin Heidegger (1962) and the idea of being thrown into the world, or as Rosen himself puts it, 'we are preceded by the unity that we try to constitute analytically as we move along within the lived present' (p. 272). This introduces the third property of the ordinary, namely comprehensiveness. The world did not begin with our experiencing of it, nor is it formed out of our experience. We were thrown into an already existing unity to which our selves and our experience are added. Ordinary experience therefore has no beginning, rather we find it fully formed: 'We are never about to enter into ordinary experience but we are always already there' (p. 272). Experience 'begins as something we are already familiar with, something in which we are already immersed' (p. 272). This 'something' is a comprehensive unity of regular experiences.

Hence we see how the ordinary rests on common sense; 'that which it is most plausible to believe with respect to everyday life' (p. 297). For Rosen, 'Common sense is thus not a body of doctrine but the ability to assess the merits of a solution to a problem on the basis of life as we currently know it' (p. 297). And so it follows that 'The structure of ordinary life is inescapable' (p. 298). We can try to deny this and act accordingly, but the consequences of doing so are inevitable. If we refuse to believe we cannot fly and put this to the test, we will most assuredly fall to the ground. We cannot deny the limits of human physiology, human nature and common sense, be it in theory building or our daily lives. What this suggests is that we have to accept the ordinary, and this is the first stage in understanding it.

Accepting the Ordinary

The understanding of the ordinary we are now developing is one based on limits. These take the form of human nature, human physiology and common sense. But this does not mean that the ordinary is synonymous with stasis or a lack of variation. As we have seen, variety is not negated by the ordinary. Rather the ordinary is where there are common rules and laws, which hold in most cases and most of the time. Moreover, we judge transgressions not on their own terms, but only insofar as they relate to the ordinary. But most actions, most of the time, are ordinary.

This can, of course, appear to be a banality. Timothy Gould (2003) writing on the work of Stanley Cavell states: 'What makes (an action) ordinary is that the action in question belongs to the ordinary world of the one who does the action' (p. 75). This as Gould admits, might not appear to be saying much. But Gould, following Cavell, suggests that there is something rather deeper here. He suggests that this definition of the ordinary should be taken to mean that 'the action does not thereby change the *limits* of the world' (p. 75). The action in question is ordinary because it does not breach the rules of common sense or seek to extend the boundaries of capability. This still leaves a considerable degree of variation, but is still crucially an act of acceptance.

It is this acceptance of the ordinary that we see continually in Cavell's work. Whilst, particularly in his later work, Cavell is also keen to consider the overcoming of the ordinary (Eldridge, 2003), he is clear that we need to begin with an acceptance of the world as it is. We can see this with the quote from Cavell discussed at the start of this chapter: he accepts that he did not make the world, but rather he testifies to the world through his existence, observation and use of language. Cavell, like Ludwig Wittgenstein, appreciated the 'sturdiness of the everyday world' (Gould, 2003, p. 77). He saw that there is a certain power held by the ordinary world. I think we can, and should, convert this notion of power to Rosen's notion of common sense, but with the important emphasis on the palpable quality that the idea of power gives to it. The rules and laws of common sense are seen as having a substance and a consequentiality. They are active elements within the world that operate regardless of our particular arguments, misgivings and attempts to halt them. We can see this as being similar to the manner in which we are helpless to alter direction when we are in the midst of a crowd all moving in one way, or the manner in which language alters through the adoption of new words and inflections. We do not know who started moving in the particular direction, or who invented the new jargon, but once they take a hold they become unstoppable.

Cavell does not argue that we should merely accept this power: he is a perfectionist rather than a fatalist. However, 'that power has to be accepted in order to be apprehended' (Gould, 2003, p. 77). We cannot simply ignore the power. Instead, if we seek to change anything, we must understand the palpable sense of the ordinary, its strength and resilience. We cannot merely wish it away and seek to build something else.

But we can take this even further. The sturdiness of the ordinary indicates two things. First, it suggests that it might actually be quite difficult to leave. As we have seen, we cannot properly operate without the ordinary, and that to be in a continual state of extraordinariness is to be in an unliveable world. The ordinary is our reference point and there is a continual pull back into it, because of the limits of our nature and reason. Second, this sturdiness indicates that the ordinary can take the strain of variation and difference. There is a resilience to the ordinary which means it can accommodate a degree of difference, so long as the common rules adhere in most cases and most of the time. This suggests that we do not need to risk all to be different, nor is difference difficult. We can accept the ordinary because, so to speak, *it is accepting of us*.

So this idea of acceptance is crucial to our discussion of the ordinary. Cavell is quite right that a first step to real change is to accept the ordinary in all its power. Its rules will apply to us whether we appreciate them or not. We cannot operate outside of the ordinary, or without reacting against it. In either case we have to contend with the limits which the ordinary establishes around us. But these limits do not oppress us. This idea of limits is, I believe, quite crucial to our understanding of the possibilities and prospects of the ordinary.

Roots limit us by preventing us from leaving: we cannot uproot and depart from our environment, or at least to do so is a considerable upheaval, which involves a risk. Such a risk may not be quantifiable in advance. In any case, to thrive again, we must be placed in similar soil. If we move from one place to another, we remain limited to a number of environments in which we can survive. I am not advocating here a form of environmental determinism, but rather that if we migrate we take much with us, and thrive where and when we find something congenial to our roots. We have to apprehend the limits this places on us.

Ruts limit the manner in which we can move from place to place. They direct us in particular ways, and these will be well-used and frequented often by travellers with the same purpose and the same destination in mind. Ruts are born out of the limitations that commonality places on us, of our wanting to repeat the same things. Again, we can strive off across unmarked ground, seeking our own way. But we take a risk and do not know what we might find. Are there things *out there* we are ignorant of which might harm us? What if we get lost and never arrive at our destination?

There are safe places and safe routes and these limit us. Yet they also protect us and increase the possibility of our surviving, thriving, being understood by others and being found by them when they want us. There is a power in limits and this is because they become known and obeyed. This acknowledgement is the necessary precursor to change, but it also shows us the utility of limits, and forces us to question why we would want to breach these limits. This is the danger of knowledge: it might lead us to a common sense.

So the manner in which we discuss the ordinary is important, but then so is the recognition of difference between an object and its representation. This difference starts to become apparent as soon as we start to wonder, as soon as we begin to open our eyes to the ordinary spectacle that is all around us. And the beauty of this is that we do not need to theorise, we merely need to look.

Framing the Ordinary

We might see the ordinary as beneath theory yet the basis for it, in that the ordinary is nothing less than the reality of our lives. And we are too busy doing it to talk about it. Philosophy, like other academic disciplines, is both too sophisticated and too narrow to deal properly with the ordinary in its full range. This may be why philosophy considers the ordinary as trivial: it cannot encapsulate it or get its measure and therefore refuses to take it seriously. We can see this problem even in Stanley Rosen's discussion of the ordinary which we relied on here. I do not think it an irrelevance nor unreasonable when I suggest that the full complexity of Rosen's critique could not be fully understood by even a well-read and intelligent reader. He writes as a philosopher and for philosophers, even as he criticises philosophers such as Heidegger, Austin and Husserl for their abstraction, sophistication and obliqueness. One could not properly follow the nuances of his argument without also having read the thinkers he criticises. But to do this would, as Rosen admits, take us away from the ordinary and into abstraction.

In answer to this criticism we could posit the entirely reasonable claim that Rosen seeks to say something that is philosophically *interesting*, that takes philosophy on by saying something that is both substantive and new. This means that any philosophical critique must itself be embedded in philosophy, making use of the apparatus at the disposal of philosophers. Unless this is done, other philosophers would rightly dismiss the critique as worthless. This means using a technical jargon, referring back to key writers' theories and arguments and exposing them to criticism, and in turn justifying one's own position.

But this does create a dilemma: how can we ever then write properly about the ordinary, in a way that fully deals with it as pretheoretical, *and in a manner that is itself pretheoretical*? Can we achieve the kind of simplicity coupled with comprehensiveness or completeness that seems incumbent on anyone attempting to give full measure to the ordinary?

Perhaps one way would be through a novel telling a story or offering a narrative of ordinary life. In this way we might describe the ordinary and its effects on us. The sort of novel that we might use for this purpose would be something like Nicholson Baker's *A Box of Matches* (2003). It deals with the ordinary life of a seemingly ordinary person, telling us his thoughts and offering some details on his life. But what is most interesting about Baker's novel is the sense of regularity and routine it has, which mirrors the routine of the narrator. Each morning he rises early, lights a fire and makes himself some coffee. And then he thinks and writes down his thoughts. Each chapter represents a new day and the novel ends when he runs out of matches. It reaches no conclusion, neither does it really have a beginning or a middle. It therefore appears to offer a way into considering the ordinary. Baker, as he has tried to do in his other novels, seeks to dwell on the minutiae of everyday life, be it feeding the family pets, filling a dishwasher or going to get a haircut. This is all done within the confines of specific situations and with no attempt to generalise. The tone of the book is conversational and whimsical, yet engaging and lacking in sentimentality.

But there is still a difficulty here in that the book is beautifully written and contains arresting images, metaphors and allusions: bullnecked and shaven-headed marines are referred to as 'penile tubes of warmongeringness' (p. 57), and a palm tree criticised as 'doing only what is absolutely necessary to do to be a tree' (p. 21). In other words, Baker's novel stands out as a distinctive piece of literature that tells us something about who and what we are. A good novel like this is engaging, fun and enjoyable, but can it therefore be said to be ordinary? Can something that stands out so much be used to demonstrate the ordinary, or is what Baker has achieved something extraordinary? And isn't this what we want in a novel? The cost of achieving the ordinary, in terms of content and style, might be to be unread, because one is not really saying anything distinctive or stimulating that engages with readers.

Perhaps what is needed here is to differentiate between the ordinary and depictions of the ordinary. Baker depicts the ordinary, but in a manner that makes us think about it, to see the similarities with our own lives. We appreciate, because of his imagery and his whimsy, the joys of the ordinary. The acuity of Baker's descriptions helps us to see the very complexity of our usual activities, and how they can be things both of beauty and interest. What Baker is able to put across so well is the distinctiveness of the regular and the familiar. The novel describes situations, thoughts and activities that are all of a type and which we are therefore able to recognise. Yet they are still distinctive, in that they belong to the narrator and his family. What we have then is a very acute description of dwelling as both ubiquitous and unique through an extraordinary description of the ordinary. The novel shows that there can be peculiarity in the common things that we do.

We can see some similarities here with the so called 'snapshot aesthetic' of photographers such as William Eggleston and Garry Winogrand. When asked what motivated him, Winogrand answered simply that he wanted to see what something looked like photographed (Malcolm, 1997). What motivated him was the creation of an image. A Winogrand photograph of a street scene or an Eggleston image of an orange truck in a field, or a tricycle in front of a house is interesting precisely because it is a photograph. It becomes distinctive, and worthy of comment, when it has been framed.

What both Winogrand and Eggleston have attempted – and achieved – is the framing of the ordinary. It now becomes an exhibit, distinct from what it hitherto was or appeared as. It becomes a *story of ...* or a *description of ...* rather than the thing itself. This may mean it is no longer ordinary as such, but without this artifice we can make no progress – we cannot see the ordinary for what it is. And perhaps we are better able to see it now it has been framed. But the image or description is still clearly connected and remains in contact with the object. The image or description has some connection because we can readily associate it with the object – it is clearly an image or description *of* something. We recognise it because it is still ordinary, even as it stands out as something distinct, even as it has become an image or a description.

But still photography can only take us so far. The ordinary is not stationary, even as we might wish it to remain still. The frame moves and we move within it. This immediately brings to mind the moving images of the cinema. Both media have the ability to stop time, and photography would appear to be better able to achieve this sense for us. After all, the photograph has caught a tiny fraction of time and holds it fast. But this is also its problem. What we seek to do is not to stop time for its own sake, but to *locate* it, and so show how it is linked to memories, emotions or actions. The best photographers, such as Henri Cartier-Bresson, William Eggleston and Martin Parr can do this, but their work does not allow for any development. We can imagine the situation within a photograph and how it might develop, but we cannot *see* this development. Photography cannot *manipulate* time for the purposes of story telling; it must merely rely on the trapping of the present and the fact that this capture frames it in a sufficiently significant manner to tell its story. But by putting a series of frames together we can develop this trapping of time, and see its fullness. So just as our frame moves as we do, and we move within it, so we can use the moving frames of film – which show actors moving – to help us see the ordinary around us.

What I want to suggest is that when we write about or picture the ordinary, what we are doing is framing it. Not only do we see it and note its quotidian beauty, but we now, as it were, capture it, or part of it, as a monument. This does not destroy the ordinary, even if it does make us see it differently.

Of course, what we really have are frames within frames: the ordinary frames us, it is what gives our life its meaning, and gives us the space to travel from one thing to another. So when we seek to frame the ordinary in our writings and our pictures, we are merely seeking to replicate the manner in which we are held. We know that the act of framing can distort or mislead: we may leave out something important from the frame, or pose the picture to emphasise one aspect to the detriment of another. Cinema, like photography, is a technology that creates a picture chemically or digitally, and which can be manipulated accordingly. It can exclude some parts of the scene, whilst dwelling on others. We need then to be guarded about its utility for our purposes, and not to make too categorical claims on its behalf. But, at the same time, we should not fight shy of using those materials which are best suited to helping us. In the rest of this chapter I wish to consider the notion of framing in more detail, and, in doing so, justify the approach I have taken in the chapters which follow.

Time and Space

The house where I live is, of course, one of type. It is one of many similar houses, an ordinary brick box made special and particular by the habitation of me and mine – by those I care for and wish to share my life with, and for whom I have an inviolable obligation.

I leave my house, closing the door to keep all else out as I go, and I enter my neighbourhood. This is made up of the special places and particular dwellings of other people, some of whom I know and speak to, and others of whom I have little or no knowledge. I walk on the pavement, one like any other on thousands of estates in Britain, showing signs of repairs and cable laying. I ignore the cats that eye me suspiciously. There is a similarity to these houses around me, as one would expect. Some are in terraces of three or four, some are semis and others detached; most in a yellow brick, but some in red, and some with a partial screed; some have their original wooden windows, whilst others now boast new UPVC in brown or white. Despite these differences, the effect is one of neatness and order, of houses and gardens well kept by people who enjoy living where they do.

There are some walks I take quite often, along routes that I now know well. I traverse across roads, along pavements, past cars, bikes, animals and people, some of whom acknowledge me and others not. As I walk I see many things, all of them, *in their way*, ordinary. But, even on these routes I know so well, I often see something that interests me and makes me stop to investigate. By slowing down we begin to see the world in its regularities and differences, its particularities and confusions, and its patterns and dissonances. Many a time I have been caught by the sight of a single red leaf in a sea of green, or the shape of a tree against the sky, or the shadows made by leaves on a tree trunk.

But all this we see is *just there*. It is merely out there if only we care to see it and take it in, to go beyond the cluttered, peremptory, distracted view of our busy lives, and if we are able to breach our ignorance as we tramp through the ordinary world, seeing no further than our feet and not getting beyond our own petty preoccupations.

The seeing of things develops the more we feel able to look, and the more we train ourselves, the more we see. But this is not a question of volume or quantity, but of depth and resolution. Look at a house and at first we might just see it entire, as a complete object with its own shape, colour and texture as it stands against the earth and sky. But then we begin to see its parts: the regularities of brickwork, and then where the patterns change, evidence of spalling, chipping, weathered pointing. We start to see particularities of colour and shape in the house, which might strike us as pleasing or not, as unusual. We begin to see the house as distinctive, as showing signs of personalisation in the garden, the paintwork, the curtains, the glimpses of ornaments and décor. This may attract us or repel us, and, of course, it may give us an entirely false picture.

But what we initially saw as an ordinary house – and what is an ordinary house – takes on a distinctive quality, even when we find the decoration not to our taste. What we begin to see is its particularity, that, whatever we may think, it is the *only one* for those who chose it, and chose to make it as they have, as a place where they can love, care and share with those they are obliged to. We need not know them, or care for them. We need never even see who it is that lives there. And nor need we remember what we have seen for any length of time. What matters here is that things that appear the same, and which most determinedly are the same, and which revel in the anonymity of sameness, are particular and present a special case for those to whom it matters. What we need to do is to see that and only that; not to pry and peek into the lives of others, but to see the particular in the lives of those people it frames. The framing makes it distinctive.

We, of course, do not want others to stare at us and ours. The anonymity of our dwelling is a defence and we might bristle at those who try to observe us too assiduously; we feel wary of their interest. I want to take my particular dwelling for granted, so I can get on and do what I wish to do. I seek anonymity then, but in my unique and distinctive ways. Anonymity gives me the space to be particular, and anonymity comes from sameness. What this means is that this distinctiveness we have in our situation is a cognitive and not a material aspect of how we live. It is how our lives are centred, how the frame fits round us, that matters here.

The key question when looking at the ordinary is 'can something remain ordinary and be noticed as distinctive?' So when we look at a still or moving picture of a house – which consists of the normal elements of a house: walls, doors, windows, roof, and so on – what makes it interesting to us? And does the engagement with this image stop us seeing the house for what it is – an ordinary place where people live? We observe the house, we take it in as an object that has been photographed, and view it as a distinct entity, so that it becomes, so to speak, distinctive.

One possible answer to this dilemma, as we have seen, is that we see the object as an exhibit of ordinary things. If we take the example of a photograph of a tree, the first impression we might have is of a confused mass of twisted limbs and leaves. There is no pattern, form or shape here, merely a discordant mass. Yet as we look more closely we see interesting shapes and patterns, and our eyes are drawn into the photograph. The tree then begins to gather a shape and a depth, and we disentangle the many branches and blocks of light and shade. The more we look at them and allow our eyes to be led by the complexities of form, colour and contrast, the more we see a grandeur *and* a particularity. To return to our discussion in Chapter 1, we see that there is no such thing as *the* arborescent system, only different and distinctive trees with an individuality of their own. They are every bit as complex as any rhizomic structure, and as unpredictable, yet this comes from a sense of an ordered whole and is not reducible to its constituent parts.

What this suggests is that the more we look, the more we will see, and we will be drawn in beyond the surface appearance and into the complexity and grandeur of the ordinary. But we also see that this complexity and grandeur derive from simplicity. It comes from a few basic elements and these are commonly around us. We do not have to search far to find these things, but rather we just need to look and allow ourselves to be drawn in. And this, in turn, draws attention to what grows around us, so that perhaps the next time we see a tree in blossom or a maple swaying in the winds of autumn, we might stop and wonder.

Of course, this might be a temporary revelation. We soon move on and start to look elsewhere. The blossom drops from the tree and gets trampled underfoot. We might not be able to recall exactly which tree it was we stopped to admire. Yet we are somehow changed by these ordinary acts. We do not entirely lose the revelation, for when we next look at a tree we know that it has component parts and that these have some independence and difference from the mass that we instantly observe. And we begin to comprehend the complexity, the wonder of the thing.

This is very much what Andrei Tarkovsky seeks to achieve in his film making. In films like *Stalker, Solaris* and *Nostalghia*, we see him dwelling on apparently banal views of nature. His camera scans across a pond and dwells on the plants moving slowly backwards and forwards just under the surface, or on a jumble of apparently random items in a stagnant pool. At first we see nothing of significance, but as the camera stops we pick out particular things – a syringe, a torn page from the Scriptures, some coins – and as we begin to piece these things together they suggest other motifs and images to us. The more the camera dwells, the more we can see, and the more we look, the greater is the significance of what is before us.

There is an order to the ordinary, which we can apprehend. The films of Tarkovsky have a depth that can only be appreciated with effort, through a concentration on the images until things start to appear, to show themselves from the depths of the scene. This gives us a range of perception and an understanding of the complexity in ordinariness that we can find in nature. We must not expect the natural world to do our work for us.

Over the last century artists have increasingly made use of ordinary objects and have sought to create art out of anything. Arthur Danto, in his book *The Transfiguration of the Commonplace* (1981), seeks to deal with the trend in modern art to use ordinary objects and to transform these into art. The obvious examples of this are Marcel Duchamp's urinal and Andy Warhol's use of soup cans. Danto discusses the transfiguration of the commonplace, of taking ordinary objects and turning them into high art, and how the artist, as it were, 'exceptionalises' the thing into art. Danto considers this to be a considerable shift in the nature of art, which for him justifies a new philosophy of art. This philosophy might at first appearance seem appealing to a consideration of the ordinary or commonplace. Yet in fact it is the very opposite of the position I wish to take. The acts of creation of Duchamp and Warhol are to decontextualise the ordinary object; the object is taken out of its proper and usual place by the artist.

It is this sense of context that I see as being so crucial to the representation of the ordinary, and this returns us to our consideration of Mikhail Bakhtin's notion of the chronotope. As we saw in the introduction, this is his linkage of space and time. To reiterate, Bakhtin defines this concept as 'the intrinsic connectedness of temporal and spatial relationships that are artistically expressed in literature' (Bakhtin, 1981, p. 84). So, if we wish to understand the motivations of a person, we need to locate them to a specific place, and this person cannot be fully comprehended until we appreciate their linkage with a place. This is to suggest that time is stopped and put into a particular place. It is this which film is capable of achieving, and hence artists like Tarkovsky have used Bakhtin's chronotope to assist them in their linking of memories and places. We can see *The Mirror* as Tarkovsky's chronotopical signature, as his attempt to link his own self to those places he was found in. These are the places of childhood, when he had an innocent directness to the people and things around him. In films such as *The Mirror* time is stopped and is localised.

Film has a further quality that is important for those of us wishing to enhance our understanding of dwelling. It has the ability to use place to interiorise. It can be used to show emotions and to link them to a particular context. Places can, as it were, act as a mirror to our feelings: they can set the mood by indicating feelings such as fear (for example, *Panic Room*, see King, 2004a) and loss (*Three Colours: Blue*, see King, 2004b). We can connect this again to Bakhtin who suggested that time could be 'full', and that an artist could make use of the pregnant possibilities offered by linking time and place. What this means is that time, when linked to a specific place, connects the past and future with the present. Natasha Synessios (2001) suggests that Tarkovsky captured this sense of the fullness of time in *The Mirror*, where he uses recollection and memory, dreams and visions along with contemporary dialogue. He uses time to the full, so that it includes all potentialities and maximises the emotional and spiritual symbolism for the viewer.

There is a ready link here with the manner in which we use and relate to our housing. Our dwelling has a timeless quality for us, or rather it is full of time. Our dwelling has no start or end for us, but instead seems to form a seamless continuum of presents linked to the past and future. It holds our memories and our hopes and dreams for the future, all along with our present activities.

Our dwelling is the specific relation of time and place. This is what is meant by the concept of *protected intimacy* as the key element in our dwelling (King, 2004b, and see Chapter 4). We dwell in a particular place, which is shared with other people and things, full of all our memories and hopes, and where we are free to love, care and share without unwanted intervention and observation. In this place, where our intimacies are protected, space and time are intrinsically linked. This place is a store of memory, a refuge for all our hopes and dreams, and a place to avoid our fears.

Finally, using film allows us to place the artifice involved in actually seeing the ordinary. We almost need to take ourselves out of our ordinary – to be outside of our place – for us to see the ordinary as such. Film helps us to do this, because of its linking of space and time, and the manner in which ordinary spaces are granted significance by their linkage with action in time. A film is an artifice, and it is therefore not real life. So when we see the ordinary in film it is not *really there*. But by the judicious use of examples, we can suggest something of the ordinary and show how it and housing are linked through the intrinsic connection of space and time.

This discussion of photography and film, of course, does not give us a definition of the ordinary, any more than our discussions of the work of Cavell and Rosen did. Yet what we now have is an understanding of some of the properties of the ordinary and how these properties can be exhibited. When we start to see what we are in the midst of, we can frame it: the act of seeing is to define the ordinary as particular. It becomes space with a purpose, because it is linked to a specific time. We might suggest that this framing sensitises the place so that we recognise its virtues. The act of framing turns it from an anonymous space into a place with meaning. And we are able to glean this meaning as we take our time within space and learn to accept it, just as it accepts us.

What I wish to suggest is that once we accept this space as our ordinary environment, we become part of it and we cannot separate ourselves from it. We too are absorbed into the unity of the ordinary, and its comprehensiveness encloses us and our experiences. The space becomes sensitive to us and us to it: it becomes our background, the stage upon which we can play out our ordinary lives.

Chapter 4

Housing in the Background

A house is a static object; but the people who use it are not. They move around and use the house. They take it as space in which they can act and be. There is, then, mobility within a dwelling. And the house, as the place in which we dwell, sits behind that action. It enables the agency and accommodates it. The house accepts us as we are.

This is what the ordinary means and this is how we should see it: it is a facilitating space, a space to hold the actions within it. In this sense, housing has an incidental quality, which can be seen as the ontological corollary of housing as a means rather than an end in itself (King, 2003, 2004b). It is to see housing as the background – the stage – that allows us to act. It is the set, the locale, where we play out our lives with other actors and their sets overlapping and interlinking with us.

But we do not go to the theatre or the cinema to look at the stage or the sets. Doubtless, we want these to impress and we know that a good set can make or break a play; and we want a film to look good and the setting to be appropriate to the action. But we are interested in the forefront and not the background: we are concerned with what is moving about in front of the set. And if this works we forget that we are in a theatre watching an artifice. Instead, we have the illusion of being with the actors, of being taken in by them into their world, so that it stops being a mere set on a stage or a film studio. Likewise, when we are in our dwelling we know we are in the midst of something supportive and which holds us, allowing us to lead our lives. But what matters is that we can play out our lives, not that something holds us. And so we can take the receptacle for granted and use it to further our purposes.

I want to show that this does not in any sense diminish the role of dwelling. In fact it is intended to do the reverse: it operates with such a facility, so smoothly, that we do not have to take our eyes away from what we are doing. This is like any tool operating, to use Martin Heidegger's jargon, as ready-to-hand (Heidegger, 1962). The tool appears to be an extension of ourselves, operating in concert with us. This phenomenological state continues so long as the tool continues to work in its proper manner.

There is a danger that this taken-for-granted nature of dwelling can lead to a forgetting of its virtues, and hence we try to 'improve' on what we already have in the belief that we can make ourselves and our 'equipment' somehow more distinctive. In Chapter 6, I shall consider the dangers and problems of this fallacy, but here I wish to consider the *ordinary* role of housing. I want to relate housing more directly to the ordinary and to explore the epistemological qualities of housing. I use films to help expand my discussion, at least in the first part of this chapter when I explore the incidental quality of housing. I shall offer some examples of the manner in which dwelling is used as a background to create a particular mood or to tell us something of the psychology of the characters. In the second part of the chapter I shall develop a picture of the epistemology of housing, as the embodiment of roots and ruts. This draws on some of my previous work on the social philosophy of housing (King, 1998, 2000, 2003), and extends it by linking it to my concerns for the ordinary considered in the previous two chapters. In this part, I shall question the assumption that the principle subject of housing research is housing policy, and instead propose a focus on housing as an activity undertaken by households and individuals. This debate is partly a question of scale, about whether housing should be seen as the preserve of government, planning agencies and large corporate landlords, or as an activity we all take part in as individuals and households. On one level this ought not to be a debate that we need to have. The nature of housing markets and the manner in which housing systems are organised in nearly all societies should lead us rapidly to the conclusion that housing is something that households do, on occasion with the assistance of government, but sometimes despite official hindrance. Yet I fear that for most housing experts this idea will seem preposterous. The notion that housing is about policy making, planning and setting targets is so ingrained that most housing experts have lost sight of how they live themselves. But when we ask ourselves how much of the lives of our fellow citizens is made by housing policy we see it is rather little. And where this is the case, say in the social housing sector, how many of these tenants would see housing policy as being the determining factor in their lives? The main 'housing relationship' they have, if we can talk of such a thing, is with the particular dwelling they occupy. This is a personal and private relation in which the landlord is a useful adjunct, but more often someone that can be safely ignored.

The really significant point I am trying to make here is not some claim for economic or political liberalism, but rather the epistemological status of housing. Our dwelling forms part of our ordinary world: it is part of what we are in the midst of. We take it for granted in the knowledge (or hope) that it works and will continue to do so without any external intervention or even particular attention from ourselves. So regardless of the level of intervention by landlords, planners or government, regardless of changing policies, we are capable of relating to our housing in a personal manner that sees it as our common place: the place where we are free to act and on the stage of our own making.

Housing as a Stage

Where films tend to be about houses or housing they, perhaps naturally, tend to be dramatic. Perhaps the most notable (and only?) film about housing policy is Ken Loach's *Cathy Come Home* (1966), which shows the degrading consequences of homelessness and bureaucratic insensitivity. When we look at houses instead of policy, the results are equally dramatic, for example David Fincher's *Panic Room*

(2001), which depicts a woman and her daughter trapped in a house by three burglars (King, 2004a).

Perhaps the most common examples of the use of houses in the history of cinema is the haunted house, a trope discussed by Anthony Vidler (1994) in his study of the architectural uncanny. Following Freud, Vidler sees the uncanny as the familiar which returns to us in an unusual or abnormal manner. The haunted house is one such familiar object that turns out to be unhomely.[1] This shift from the familiar to the unknown is significant in that it shows how a taken-for-granted object can quickly turn to horror, be it Tobe Hooper and Stephen Spielberg's *Poltergeist* (1982) or Alejandro Amenábar's *The Others* (2001).

Whilst this is an important use of housing, I wish to concentrate on examples where housing is used, but not as a dramatising element. Rather I want to consider the opposite of this, where housing is in the background, and is used as contrast or to add subtlety to a particular portrayal or situation. I wish to dwell on those examples where housing is in the background and thus it is understated. I want to show how housing resonates when it is *heimlich*, or homelike, rather than when it is uncanny. Housing is what we go home to when we are finished, and so it is appropriate to see it as familiar, and as background, where the story is not *of* housing, but *in* housing; and where the story is not told through housing, but where the dwelling merely plays a part. We might say that the house is a character, not in the forthright sense of the haunted house, but as a minor player whom we see initially as not essential to the plot, but who sets a scene up, provides the mood or qualifies a particular condition. It is where the dwelling is *additive* but not determining. In this manner, housing in film is as housing is within our lives: it forms an ordinary part of our lives that tells others something of us and sets the scene up in some way, but is not the reason for that scene itself. Housing, then, is in the background.

One particular example of this, and precisely because it uses an ordinary domestic setting to offset the supernatural, is M. Night Shyamalan's *Unbreakable* (2000). This film concerns an apparently ordinary man, David Dunn (played by Bruce Willis), coming to terms with the fact that he is invulnerable. He lives in a typical home with his wife and son. His marriage is faltering and he and his wife are considering separating. Their son is desperate to keep them together and needs his father to be someone special. The first time we see Dunn is on a train returning to Philadelphia after a job interview in New York. The train crashes and Dunn is miraculously the sole survivor. The rest of the film is about his increasing awareness of his supernatural invulnerability and the uses to which it can be put. In this he is continually pushed by an exotic character, Elijah Price (played by Samuel L. Jackson), who is convinced of Dunn's powers.

Shyamalan, as in all his films,[2] uses colour symbolically and to show particular moods. Price dresses in purple to show off his exoticism, whilst Dunn's clothes are nondescript. The colours of Dunn's family home are dull greens and browns, which emphasise their ordinariness. There is no hint of design in this

---

[1] The literal translation of the German for uncanny – unheimlich – is, of course, 'unhomelike'.
[2] See, in particular his use of red as 'bad' and yellow as 'good' in *The Village* (2004).

house, only practicality and normality. The only bright thing we see in the house is an orange note stuck to the wall with the time and number of Dunn's train. This contrasts so markedly from the prevailing décor that we cannot avoid its significance. The colour orange again denotes danger later in the film, when it is used for the overalls of the shopping mall maintenance man whom Dunn follows and who we find is a murderer and kidnapper. So the colour orange is used to alert us to an imminent danger. But in contrast, the Dunns' home contains nothing special, nothing grabs the eye, and thus it operates as the opposite of the supernatural. It is this taken-for-grantedness and obviousness of the domestic which can so readily be used as background. The home is used precisely because it is nothing special. But this ordinariness is essential to the development of the plot, which hinges on the distinction between the normal and the abnormal, and the means by which we come to notice it.

A further example to contrast the normal with the abnormal – or more properly appearance with reality – is in the Wachowski Brothers' *The Matrix* (1999), in particular in the early part of the film where we are shown Thomas Anderson (played by Keanu Reeves) in his dishevelled room. In these brief scenes, and others in Anderson's workplace, the directors are keen to show the weakness of this character and his apparent unpreparedness to fulfil what fate has in store for him. In addition, these scenes in his room also demonstrate his solitude. What is intended here is a demonstration of normality, a linking with the audience's concerns and means of existing (who, after all, are the very people who are supposedly obsessed with *The Matrix*, but computer nerds and geeks who exist through computer games, kung-fu magazines and anime films?). Anderson is unable to make his bed, tidy his room, or get to work on time, yet he is the man destined to save mankind from their computer-generated prison. These scenes of incompetent domesticity contrast with the special and heroic qualities necessary for the progression of the action within the film. But the state of Anderson's room also shows he is distracted and driven in his search for something beyond his normal life. In no sense is this film about housing: any such belief would be absurd. The point is rather that the directors are creating a mood, and making a comment on the character of their hero by their use of dwelling, which is then contrasted with the action and fantasy scenes later in the film.

This characterisation is not particularly subtle, but what it shows is that we have some common notion of what housing is like, and directors such as Shyamalan and the Wachowski brothers can play on this. There are certain tropes which resonate with us because we understand their significance and what it means. We take certain meanings from an untidy and unkempt room, and different ones when we see space that is ordered and regular. A rather different example of this apparent incidental quality, which uses a dwelling to describe the psychology of the characters, is Ingmar Bergman's *Summer with Monika* (1953). This is the story of two young lovers, Monika and Harry who, after Monika has left home and Harry has lost his job, go off in a motor boat to the islands off the coast of Sweden for the summer. They are forced to return as summer ends and because Monika is pregnant. Their idyll does not last long into their marriage with Monika regretting the confinement of domesticity and motherhood, whilst Harry finds a good job and begins studying to be an engineer. Monika cannot accept her situation, whilst Harry can, and so the film ends with him literally holding the baby as Monika leaves.

The film is one of contrasts: between freedom from responsibility and the drudgery of commitment; between ideals and reality; and between order and mess. All of these contrasts are formed through the portrayal of domesticity. We see the contrast between the orderly nature of Harry's aunt, who keeps Harry's father's house tidy, and tries to do the same later for Harry and Monika's flat, and the disorder of Monika's parents, with their rather squalid, cramped and noisy flat. Bergman also uses a disordered domestic environment to show Monika's capabilities and interest in being a mother and housekeeper. What this disorder is meant to register is first, a lack of competence; but second, it also represents a lack of commitment. Monika is not prepared to accept that care is a permanent obligation and not merely a summer fancy: one must remain committed to specific others – husband and child – and to the broader society and its morals. We can suggest these are represented by Harry's aunt, with her sense of duty and civic responsibility shown in the scene with the registrar, where the aunt convinces him to let the two young people marry.

But what takes Bergman's film beyond the quotidian, and which makes it a genuine cinematic statement, is that it also considers the idea of breaking out of domesticity. The two young lovers see no need for a proper dwelling, spending the summer on a boat. Yet this attempt is doomed, and we know it is as soon as Monika announces she is pregnant. At this point responsibility starts, and the couple have to return to domesticity: parenthood and climate make this inevitable. They seek to impose an order on their lives, to which Harry adapts, but Monika cannot. At the end of the film is one of Bergman's greatest filmic gestures, when he has Monika (played by Harriett Andersen) challenge us directly, with a prolonged and provocative stare straight into the camera, daring us to criticise her, but almost forcing us to look away. This one scene transforms the entire film: prior to this moment we have seen Monika as flighty and perhaps insubstantial. But with this prolonged look she takes the initiative away from the viewer and challenges our opinion of her.

Monika's provocative stare is contrasted with the last time we see Harry's face. We see him reflected in a window as he holds up his daughter: we see his image as moderated, as mitigated by what is around him, and by the vicissitudes of the environment, whilst Monika proclaims her freedom. As if in his mind's eye, we see Monika lying languidly on the boat and running naked to the sea. His is a look of wistful regret rather than of confrontation. We leave Harry as he walks off with his daughter, and we sense he will cope as a parent, even as his personal effects are being sifted over by bailiffs and carted off. Harry accepts the ordinary, with its swings, its opportunities and closing down of possibilities. Monika, on the other hand, cannot do this; she is always looking for the extraordinary, the next adventure, and we find this challenging and provocative, like her stare into the camera. We are left to speculate on whether Monika would succeed or ever settle, but we sense she will not.

Bergman's use of the domestic rests on the ability of his audience to notice the ordinary and the exceptional: to be able to contrast the responsibilities of life with their opposite, and to come to some judgement about the main characters from this. Yet he also challenges these assumptions, directly through the stare of Monika, and by resisting the temptation of a 'happy ending' where she sees the 'error' of her ways. Bergman manages to convince us of Monika's allure even as we see her as deceitful and neglectful. He does this in an entirely naturalistic manner that does not make use of any of his more metaphysical tricks in later films such as *The Seventh Seal* (1957) and *Persona* (1966). Much of this naturalism is in the contrasts between the various domestic settings and what they tell us of the motivations and commitments of the main players. What Bergman plays on, therefore, is a commonly held conception of 'how we live today'.

Again housing is in the background. *Summer with Monika* is ostensibly a tale of young love and its consequences. Yet it is also about commitment and acceptance. Harry finds he can accept what he has, and adapts, be it to life on a boat, to living rough and scavenging for food, or being a parent, finding a job and studying. It is Harry who breaks off from his studies to calm their daughter, whilst Monika ignores her. This is because Monika wants to be elsewhere: she is constantly looking for something other than what she has. She does not accept her role as wife and mother and cannot find accommodation in domesticity. All this is shown, I would suggest, through the manner in which Bergman sets up the physicality of his scenes – the unmade bed, the dirty clothes and the squalor of their tiny flat. So this is not a film about housing, but one that is able to use our conventionality about how we live to convey meaning to us.

A final film I wish to look at here is Carl Theodor Dreyer's film, *Ordet* (1954). This film centres on a family in rural Denmark. The family consists of Morten Borgen, the patriarch, and his three sons, Mikkel, Johannes and Anders. Mikkel, the eldest son, is married to Inger whose death and apparent resurrection are at the heart of the film. Johannes is a theology student driven mad and who now thinks he is the risen Christ. Much of the film is set within the farmhouse, and more particularly the large main room where meals are prepared and served. Much of the dialogue in the film has a formality to it, based as it is on strains within the family over faith (and its lack), love and madness. This formality is dependent on and conditioned by where it takes place. The apparent space in the dwelling expands and contracts according to the emotional ebb and flow of the film, as scenes move from small private rooms, to the large common space, and then outside to the windswept coastline.

In contrast to Bergman's *Summer with Monika*, Dreyer's *Ordet* is an obviously staged and stylised piece. Many of the scenes appear like a filmed play. There are obvious set piece dialogues and scenes which are full of symbolic and philosophical meaning. Yet the film has an intensity and a motive that is enthralling, particularly in the second half of the film which shows Inger's death in childbirth and her subsequent resurrection, apparently on the word of Johannes. The manner in which Dreyer stages the film emphasises the static quality of domesticity. There is a lack of movement, and we sense the solidity and permanence of artefacts and the spaces between them. These come to represent the spaces between the family members and with other villagers. The interior of the farmhouse is dominated by a large shared space where the family congregates for meals and to talk. We only get glimpses of other rooms. The only other rooms we enter are Anders and Johannes' shared bedroom in the opening shot of the film, and Mikkel and Inger's room on several occasions, most notably when Inger is laid out in her coffin. On this occasion the room appears much larger and is very nearly empty of furniture apart from some chairs and the open coffin. Unlike the rest of the dwelling, in this scene the predominant colour is white – the curtains, the walls, and the gown Inger is laid out in. There is a sparseness here, but also a sense of freshness and purity, given by the diffused light seen entering the windows covered with white net curtains.

Dreyer's use of space in this long final scene of the film is magnificent. It begins with openness and light, where we see the grieving family and friends congregating, and the priest saying his words over the coffin. Mikkel, who has lost his faith, is finally able to cry and show the fullness of his emotions at the loss of his wife. But as the scene develops with Johannes' entrance and his summons to Inger to rise, the space becomes ever more enclosed so that all the audience can see is Inger and Mikkel. As we are opened up to the infinite by Johannes, through what is the revealed word of God, we are shown its effects at the most personal level by this use of a close up. By the complete reduction of any background Dreyer emphasises the spiritual and the inner life which is reborn in both Inger and Mikkel, and so it is appropriate that the last words of the film are 'Life … life'.

Ordet operates, therefore, through the use of confinement of space to create a sense of meaning and to heighten the symbolism. In another scene, the conversation between Morten and the tailor, Peter Petersen over the possibility of Anders marrying Peter's daughter, Anne, takes place in a bare and spartan room in Peter's house, where there is no evidence of warmth. This is no room in which to compromise or to seek solace and sympathy from those certain of their views. Morten is from a different protestant sect to Peter and thus the match is initially declined (indeed, Morten took some persuading initially to allow Anders to pursue the suit of Anne). The principles of the two men are too rigid to admit the prospect of compromise or accommodation, and thus their children seem fated to be kept apart. This dialogue takes place within a confined space cropped by the camera, and this merely emphasises the closed minds and heavy, portentous thoughts of the religiously dogmatic. We sense the rigidity of principle and the inability of the two men to compromise or shift ground in these surroundings. Both men exude a sadness, almost as if they know the futility of a discussion where neither can give way. Of course, after the death of Inger, Peter agrees to the match. He does this amidst tragedy, but within the light, uncluttered room where Inger is laid out. This, by contrast to Peter's house, is not a space that shows dogmatism or the limiting of possibility, and this is emphasised dramatically by Johannes' words ordering Inger to rise up.

Dreyer uses a very limited space – a few interiors only – to put across complex philosophical and theological ideas on the nature of faith, tolerance and commitment. What the film demonstrates above all else, however, is the need for accommodation, and to accept the conditions and trauma of others. In doing so, we can be reconciled and see what joins us together instead of what separates us. In this sense, it would be interesting to contrast *Ordet* and *Spring and Port Wine* discussed in Chapter 1. Both films have a dominant father who rules the family through principles that appear outmoded, and from which the younger generation would like to escape but cannot. Anders wants to marry, but needs his father's approval, and the children in *Spring and Port Wine* wish to become independent, but cannot break away from home. Both sets of children appear to appreciate that they cannot force a change, as this would in some way destroy the natural order of things. Instead any change must come from within the family as an organic development of the relationships.

Both films are concerned with the rigidity of principles and the consequences of this: with what happens when abstractions instead of human emotions are enacted in human relationships. Yet both patriarchs come to see the faults of their rigidity and start to see beyond it. They do this by becoming aware of the qualities of their families, seeing them as individuals with needs and feelings, and not as the subjects for principles, which, in any case, are not of their making.

The tone of these films is, of course, markedly different: *Ordet* is dark, slow and full of symbolism, whilst *Spring and Port Wine* is whimsical and warm. Yet both show the impact of rigid principle on the lives of others, and both are about keeping a home and family together. They are concerned with the need for acceptance and how we are forced to try to achieve this if we are to survive and thrive. Accordingly, both films end with a family reunited and acting together with hope restored.

These films offer us the means by which we can make some assertions on the nature of housing and how it so closely relates to the ordinary. Like the other films we have investigated here, they place housing in the background and push human action into the foreground. When we come to films as substantial as those of Bergman and Dreyer, we see the manner in which the dwelling offers a powerful insight into the psychology of the film's characters, and how the nature of dwelling can be used as a cipher and to heighten the dramatic import of scenes. Housing is something that just *is*, but it is also a symbol of intent and recourse: it is both found and expressive of our selves.

We, of course, do not look for dramatic import in our dwelling, but we are able to rely on the familiar qualities it offers. We are safe in the knowledge that we are accepted and accommodated in and by our dwelling. It is the background to our lives; it is what we are in the midst of and we need now to try to understand what this means in more concrete terms. We thus move from a discussion of film to a more explicit discussion of housing.

The Epistemology of Dwelling

What does it mean to say housing is the background to our lives? Quite simply, it is that thing which we are in the midst of. Housing is that which we move in, which offers us protection, which allows us comfort and security. But the particular quality I wish to stress is that housing is something which is *incidental* to our lives.

Housing has the quality, so to speak, of *thereness*: it does not need reflection for it to operate as we would wish. It supports us just as it relieves us of the need for reflection.

For housing to be incidental does not mean it is not essential to us, but just that it can and does work without our having to think about it. It is always there, working away for us in the background, whilst we are pursuing our more deliberative ends. What is more, we can enjoy housing along with our main purposes. If it were not there we could not complete those purposes. We would have no comfort, privacy or security and we would be bereft. So when we are living our lives and pursuing our interests we, as it were, take our housing with us. It forms the basis upon which we can act, and this is the very reason why we are able to ignore it and take it for granted. It is fundamental to us, but also something that we need not hold at the centre of our lives. It is in this sense in particular that housing is ordinary. It is so normal, so part of what we are and do, that we are able to lose sight of it. It becomes transparent to our consciousness (Giddens, 1991). Housing as ordinary becomes the order in our lives that brings with it the freedom to act. It is only when it is absent that we see it in its full significance, and have to rely on memory to help us through our exile (as I discuss in Chapter 5).

Perhaps this incidental quality is the hardest for a housing academic to accept. To admit as much would be tantamount to claiming insignificance for oneself. We do not relish spending our life's work on something that might remain in the background, a supplement, an adjunct to the main attraction. And so we naturally seek to claim some seriousness and import to what we do. But the problem is that we place importance on the wrong elements in our search for significance. We try to see housing as something concrete and substantial, and thus linked into economic and political institutions. We make a claim for housing by seeing it in terms of policy. In this way we can ignore the rather nebulous epistemological questions I have been concerned with in this book. Policy allows us to concentrate on 'relevant' issues, relevance being determined by the closeness with which we can attach our concerns to the political process.

So what I want to distinguish between here is *housing policy* on the one hand, and *housing as ordinary* on the other. My claim is that housing as the background to our lives presents it as crucially important, and that it is the *normal condition of housing*. Housing policy is only concerned with a small area of what we might call 'housing', and the imprint that it has is rather minor and limited. And it is my view that the limited interests of housing policy do not connect with how we actually use housing.

First, it is a problem that housing policy is concerned with only a minority of households. Because policy is determined by institutions and the influence of the state, it is restricted to those housed or assisted by the state, and to those issues and practices with which the state and institutions involve themselves. These include such matters as the construction, management and maintenance of social housing, the behaviour of social housing tenants, and general questions of the planning, production, consumption, supply and demand of housing. The majority of these issues are restricted to the physical qualities of housing, and more particularly the actual fabric of the properties. These activities stop, as it were, at the front door and do not concern themselves with how the dwellings are used (King, 2004b). But where attention is being paid to what occurs behind the front door, this is rather limited. Attention is generally only paid to those households who can be effectively disciplined, and this means social housing tenants. This is either because they are restricted by specific tenancy clauses, or they or their landlord are in receipt of government subsidy. Where intervention is in theory more general, for example, in the case of England, with policies such as Supporting People and those dealing with anti-social behaviour, this tends to be in a standardised and prescriptive manner, and where the nature and level of service is determined institutionally rather than on the basis of the clients' needs.

But there is a further difficulty: housing policy is something that must be contemporary: it is always related to the current time and context. This is because policy – and the study of it – chases the immediate. Policy makers and those commentating on it are concerned with current policies and practices and how they can effect change in the present. The result is that we often only see housing in these terms, as about policy making and implementation. This, in turn, is linked to the misguided notion that housing policy is a distinctive entity, deserving of special attention. This creates the fallacy that dwellings stand alone as entities, as ends in themselves (King, 1996).

But what is missing here is the use to which dwellings, be they created by housing policy or market activity, are actually put. Housing policy is concerned with the building of a small number of them, the location of many more, and issues of standards and spending levels, but this is all (King, 1996). It might sound like a substantial list – admittedly it is a necessary one – but it does not take us too far in understanding the significance of housing. We do not suggest that the significance of the motor car is just in the numbers that are built, what colour they are and whether everyone can afford one. Nor do we suggest that we only attempt to control the behaviour of certain car drivers, whilst ignoring the rest. Yet this is what we limit ourselves to with housing. The fault is that we tend only to consider as important those issues that are driven forward by government, and forget that the policies and practices of housing organisations are not exactly at the forefront of those who live in their dwellings. The important relation is not between landlord and tenants or electors and government, but between household and house. Policy can matter, but only because of the way in which it enhances or detracts from that relationship.

Of course, there is some attempt in housing policy to connect up with those who live in the dwellings. But the effect of this is often self-defeating. We can see this by considering the jargon in which policy is concocted. The most glaring example of this jargon, of course, is the use of the term 'homes' to refer to brick boxes built by social landlords and private developers (King, 2004b). There is an apparent belief that calling dwellings 'homes' connects more with the eventual users. However, all it does is devalue the concept of home and denude it of any serious meaning. 'Home' just becomes another technical term used by professionals, such as 'dwelling unit'. This is by no means the sole example of taking ordinary words and phrases and turning them into technical terms with a specific meaning. Government appears to consider that giving its policies and institutions what might be called soft and inclusive terms somehow alters the tenor of the policy. In terms of policies introduced by the Blair government in England, we can point to the mechanism funding provision for the elderly and those with special needs banally called *Supporting People*; the policy aimed at improving housing standards in the social sector referred to as the *Decent Homes Standard*, with the consequence that a house now becomes *decent* only if it fulfils national criteria relating to such things as the age of kitchen units and boiler, and thermal insulation; and, third, the campaign against anti-social behaviour organised by the Home Office called *Together*. All these policies expropriate general expressions which can now only be used in a specific sense that empties them of any other meaning. It is now the case that a house is 'decent' because of a national standard regardless of the views of the landlord or the person living in it. More generally, a term that has connotations of politeness, respectability, conventionality, is now reduced merely to a technical term for a minimum standard. One can see the need for quality housing and the imposition of standards, but without this decline into banality.

I would argue that the reason for this use of banality is precisely because policy makers are aware of the disconnection between their policies and the manner in which housing is used. Notions of decency, togetherness and support have a natural resonance with the manner in which we live in dwelling environments and communities, and so it is hoped to gain by connecting functional policies to these terms. However, there is no change in the nature of policy making, in that these policies are national standards, assessed through top-down target setting and sanctions. All these policies have is a rather gentler, if less meaningful name – does *Together* actually tell the uninitiated anything about what it is? – but are no nearer to connecting with the manner in which we use our housing.

So what I am seeking to attend to is not the abnormal actions of housing policy. What I am interested in though is the activity of housing itself, or what I have referred to elsewhere as dwelling (King, 2004b), that state which provides 'protected intimacy' (Bachelard, 1969, p. 3). Dwelling is both a physical and an ontological condition whereby we feel secure, stable and complacent. What provides this condition is the housing itself, and this returns us to the idea of housing as the background or as incidental to our lives.

The incidental quality of housing is not the same as irrelevance. Housing is still full of meaning. However, that meaning is implicit in the use of housing and not readily expressed. When housing works it is not in the forefront of our lives. The incidental quality of housing manifests how we are able to control our immediate surroundings, how we can direct the staging around us. It remains incidental insofar as we can and do control it. A dwelling is so important precisely because it provides us with the proper stage on which we can perform. This tells us why policy can never properly connect with how we live. Policy has to make things explicit and this places our housing in the foreground. In other words, *policy makes housing problematic for us*.

What this suggests is that the ordinary, when attached to housing, becomes a matter of epistemology, a matter of understanding the manner of we connect with one of our most important taken-for-granted relations. What we need to do, in order to concretise the ordinary experience of housing as such, is to define the *epistemological qualities of housing* (King, 2003). Housing, as both a good and an activity, is something that we know a considerable amount about. We are aware of what constitutes good quality housing, and whether we are currently experiencing this level. As our needs and requirements change, we are competent to determine what must change in our own dwelling. This, of course, does not necessarily mean that these changes occur, but we are still able to point to the problems. Thus, unlike other existentially significant activities or necessary goods, we are well placed to understand what housing does for us.

In particular, I have argued that there are three qualities of housing, and these allow us to relate to it as the necessary background to our lives. First, our housing need is *permanent*, as we must always have a suitable dwelling. The actual need we have for housing does not fluctuate as our need for health care and education does. There may be specific needs which we have that alter as we age or if we have a disability, but these are in addition to the basic existential need for shelter, which is ever-present. What differs is not our housing need, but how well it is currently fulfilled. And when our housing need is not fulfilled we are aware of the nature of the problem and can readily point to the solution. All of us, even those currently homeless, are able to state what good housing is.

The second quality follows from this permanence, and is that our housing need is *predictable*, thereby allowing for a more regular pattern of provision. Our housing needs change slowly to the extent that changes can be readily accommodated. Even where there are crises, such as a structural failure, flooding, etc, the means of remedying these situations are themselves entirely predictable, and hence we are able to take the necessary precautions and insure against them. However, most changes are not so fundamental, such that the very resilience of our dwellings allows for both them and us to adapt to changes. The space within our dwellings is adaptable, and the very protective carapace they form over us allows us to meet changes with some sense of security. We have grown up in and with housing of this type, and we know it for what it is. Hence we can adapt and know it will adapt with us. So not only is the need for housing predictable, but also the means of its fulfilment.

As a consequence, the third quality becomes apparent: housing, because of its permanence and its predictability, is more readily *understandable* in that we know we need it, that we will always need it, and to what standard we require it. We know what housing is and does in our lives, and we know what its absence would imply and how this would limit us. It is this ability to understand what our housing is and does that constitutes it as background. As we are able to form housing's role cognitively, we can, as it were, forget it. Instead of our needing to hold it in the forefront of our consciousness, as we would if we had the unpredictability of an as-yet-undiagnosed illness, we are able to put our housing behind us and concentrate on what matters to us now.

These three qualities are of fundamental importance to any appreciation of housing, as well as showing us precisely why housing policies need not impinge upon us. Much of the complexity of housing is a very condition of its supply, particularly the regulation of that supply, and hence the outcome of policy itself. But in terms of our experiences, we know sufficient to be confident and complacent about the relationship between ourselves and our housing. Clearly there are implications here on several levels: for specific policies themselves; for the general nature of policy making about housing; but also for the nature of housing as our common place.

It is this last sense that I am interested in here. The three epistemological qualities allow us to see housing as what we are in the midst of. Housing, because of these conditions, can be the stable platform and the secure anchor. We need not, and do not associate housing with trauma. And where we are facing trauma, be it of loss or exile, we can take solace from memory and of time stopped (see Chapter 5). We can return to the solidity of that housing which has formed us. These three qualities inform us of the resilience of housing, be it in memory or in physical space. This, in turn, points to housing as a common place, both in the sense of always being there – our commonplace – but also as something which we hold in common. The permanence of housing roots us into a cultural memory. More generally, as we saw in our discussion of Bergman, Dreyer, Shyamalan and the Wachowski brothers above, we can readily appreciate how others live, and this is precisely because of how *we* live. We share roots and ruts together and know that what makes our housing ordinary is that we share it in common with others: we are, despite the privacy we crave, in the ordinary together. But the nature of the ordinary is such that we do not need to pry, nor, as we shall see in Chapter 6, do we need to follow fashion and make our dwellings alike or according to any expert prescription.

So, to conclude this particular discussion, to suggest that housing is in the background is to say five things. The first thing to state is that we can control housing because we can understand it, it is a predictable entity, and will be needed permanently. This places us in a much more certain position than with many other goods, and certainly more so than with any other good so crucial to our welfare. Second, the relationship between our housing and ourselves, being permanent and so readily understood by us, need not hold our attention: merely to have it is enough for us to be certain. Third, these three qualities of permanence, predictability and understandability suggest that our relationship with our housing is sufficiently resilient, both as a personal and a cultural memory, to hold us and offer succour. Fourth, it helps us to understand that the ordinary is within us, and that it is part of us and is driven by us. The ordinary is both epistemologically and ontologically bound into us. Fifth, and finally, we can see that the three qualities relate closely to Stanley Rosen's three properties of the ordinary (Rosen, 2002). He described these properties as *regularity, unity* and *comprehensiveness*, and we can see that these are suggestive of permanence, predictability and understandability. Of course, the match is not on the level of cognates, and so the qualities and properties are by no means interchangeable. They do, however, relate in the sense of offering a sense of pervasiveness, stability and certainty. What they both call to mind is the coherence of the tree, with its strong root capable of offering sustenance and support, a definitive and concise shape, and a sense of purpose and clarity. Furthermore, we sense here the ideal of defined and safe ruts, clearly marked and offering us certainty. We are given the sense of knowing where we are, that we can stay as long as we wish to, and that there are clear routes taking us away and back. And because of this clarity and certainty, we can pursue our ends without concern for undue maintenance of our ordinary world. Thankfully, we know it is just there.

Chapter 5

Memory and Exile

Paul Oliver in his book, *Dwellings* (2003), discusses the distinctive architecture of the village of Akyazi in the Black Sea region of Turkey. He suggests that whilst there is a considerable variety amongst the dwellings of the village, they are all different from the vernacular architecture of the region. Instead of the typical style of a timber frame with adobe infill, the houses in Akyazi are constructed on massive posts driven into the ground. They also feature a deep veranda structure, which is again untypical of other villages in the region. But what Oliver sees as being particularly significant about this village, featuring in a chapter of his book called 'Settling Down', is:

> The residents' pride in their Georgian ancestry, their defence of their culture and reluctance to relinquish the roots of their past, and the slow, century-old process of adjustment, which includes the adoption of the Turkish language (2003, p. 55).

Akyazi is a village of Muslims of Georgian descent who originally settled in Turkey in the 1880s following the Russian defeat of the Ottomans and the reinstating of Christianity in Georgia. Akyazi is therefore the result of migration, but can now be seen as an act of remembrance, as a form of staying still even in an unfamiliar environment:

> Its houses remain Georgian in design, construction and use, and though four or five generations have passed since its birth, its present builders cling tenaciously to their traditions, while erecting dwellings that are still varied within its norms (2003, p. 55).

Despite over a hundred years of living in Turkey and the passage of generations, the villagers of Akyazi cling to their Georgian roots, even as they live and thrive within their adopted region.

When we consider the vernacular we normally see it as being where local people use local resources for their own benefit, making the best of what is available and, in so doing, adapting to the environment. Yet what the example of Akyazi shows is that we need to feel we belong, and that there are occasions when we find it hard; that because we are displaced we have lost those resources we are accustomed to, and must remake our place anew in an environment that may be hostile and uncongenial. We must still use the resources available to us, and now we call on one in particular. What can assist here is *memory*. We can aid the process of belonging, not merely through integration or assimilation, but through acts of remembrance where we insert the cues for memories into our environment.

We maintain ourselves, like the people of Akyazi, by creating a new sense of place out of the old.

In chapter two we have looked at what it means to be *out of the ordinary*, and indeed whether it is ever possible to be so. I suggested that we can carry some elements of the ordinary around with us, so that we might have an internal sense of the ordinary which would help the adventurous to survive and thrive. But I also suggested that there is a double meaning to this phrase: in one sense we can see the term as referring to 'being outside of', where we are not part of the ordinary. This may be where a person is particularly eccentric, or where we are in an environment that is so unlike what we are habitually in the midst of, where we are beyond the mundane. Yet it can also be used to refer to the fact that many of our eccentricities and foibles come, as it were, 'out of' the ordinary. These actions derive from our ordinary environment, which sponsors this action, supports it and nurtures it. We could not act the way we do were it not for the ordinary. Our actions, we might suggest, are the result of the ordinary. These two senses might not be opposites, but rather we might see the second as an explanation of the first: we are able to exist outside of the ordinary precisely because of the ordinary sense that has, or had, surrounded us. Hence the villagers of Akyazi have created a new ordinary out of the memory of their Georgian past and thus maintained the link.

Having considered the ordinary in its fullness and linked it to our understanding of housing, I want to speak of where the ordinary reaches its limits, and therefore to extend this discussion itself to its limits. I wish to consider what is *beyond* the ordinary, to see how we try to maintain ourselves in an alien environment, and how, inevitably, there are times when we are just too far out to connect to the ordinary. Such a study is partially a return to our consideration of just what the ordinary is. In particular, it is to question the notion of the ordinary as a physical space that we are in the midst of. Can we be in the midst of something non-physical? Can we not be in the midst of a mental space that we take – or rather would love to take – as ordinary?

Accordingly, I wish to consider how the ordinary relates to memory, to that sense of the ordinary we carry within us. For most of us, this might be a latent state, where what we have within us need never be manifested in any way other than patterned behaviour. This is because there is no difference between our mental ordinary and the physical state around us: it is one and the same. The question of memory becomes an issue when we are separated from this physical state – we are not where we want to be – and what we have left is only the mental sense of the ordinary. What we must now do is to try to impose these patterns onto the environment around us, just as in Akyazi, where the traditional Georgian building patterns are now marked on Turkish soil.

What I wish to consider in this chapter is where we have to work hard to find the ordinary, or where, to state the case differently, the ordinary has to work harder for us. There are some places we find ourselves in where we cannot accept where we are and what we have. In these places we are, properly speaking, lost. This might be in the literal sense of being in a formless environment, but it also might be a place where we are lost emotionally and morally. In these places, as in Akyazi, we can only accept where we are if we are able to bring enough with us. All we have here is memory, where we can look back into the past for our survival, try to use what is there, and somehow adopt and adapt what is around us now. We see this in the work of the Russian film director, Andrei Tarkovsky, particularly in his films *Solaris* (1972), *Stalker* (1979), *The Mirror* (1974) and *Nostalghia* (1983). The first, and longest, part of this chapter will therefore consist of a study of Tarkovsky's work to see how far it shows the linkages between memory and the ordinary.

But there are other places where we can do nothing but seek for an exit, and to try to return to a world in which the ordinary is or can be formed. There are some places where memory is insufficient, and where we can never be accommodated, only saved. These are places of such unremitting hostility, where we are unprepared for the conditions that face us, and where our memories of found places and of shared things can be of no comfort. Again I wish to use film to show this notion, on this occasion the works of the Hungarian director, Bela Tarr, and the American filmmaker, Gus Van Sant, whose film *Gerry* (2002) in particular demonstrates this inability to adapt and find accommodation in an environment which will not accept us. The second part of this chapter is concerned with how Tarr and Van Sant show the effects of a complete alienation with the surrounding environment that is just unbridgeable. Their main means of showing this is through the use of stillness and movement, which point to the transitory nature of our stay within the environment even as we strive to leave our mark. But even as we strive, all we have at our disposal are the ordinary virtues and vices which go to frame our lives.

All of the works of Tarkovsky, Tarr and Van Sant I discuss here show the limits of transformation and point to the manner in which the ordinary is both container and contained. We can enter new places, but only if we are prepared and take with us the necessary tools and skills. Where we are underprepared and when we get lost we cannot always redeem ourselves. Thus it matters what we take with us into exile.

And it matters so because transformation is the imposition of exile: it is where we place ourselves – or find ourselves – in somewhere new and where we have to make the change. Exile thus becomes a test of the ordinary, the limit condition by which we show the true import of what we are (or were!) in the midst of. The discussion in this book so far has been a largely positive exposition of the ordinary, seeking to explain its principal virtues and how it links to the environment around us and our sense of being centred and whole. In this chapter I wish to show the importance of the ordinary negatively. My aim is to show the importance of roots and ruts by what occurs when they are not present. By showing the alienation of exile we can only further appreciate the virtue of anchors and clear routes.

Remembering our Place

What if the place that we are in the midst of is different from the physical space that we currently inhabit? What if the ordinary is located elsewhere, in another place or in a remembered past, and we now carry it within us as an image of this place. We may remember only elements of it: there may be certain objects, smells, a smile or expression, particular acts or occasions, a word, all of which come out in a manner that we cannot control or understand. Yet any of these make us feel 'at home' in a way that we cannot find in the physical space where we are now stuck. This, I would suggest, is the problem of exile, of being displaced and yet capable of remembering place: of being dislocated yet able to discern what it is that locates us. We have a great yearning, but we cannot fulfil it with anything but memory.

Yet we are always somewhere. We are ever in some place even as we wish to be transported to another. What effect does this have on us and how can we link it to the discussions on the ordinary which we have had? Instead of being located positively we now have to deal with the negative sense, where we are stuck in a place and would wish to be elsewhere.

We deal with this problem by returning – or attempting to return – to where we wish to be. Ideally this is a physical move, but where we are stuck and find no means of actual return, we must rely on memory. This may be no cure for exile, and it may not really ease the longing, but it is a key element of the ordinary. It is one that may be largely dormant in those of us who are comfortable in our own places, but it is still necessary to understand this sense of being placed. Places are for storage and thus we need to understand how we can access what we might see as an archive of memory.

We can see memory as the ordinary absorbed, where we have taken it into us. Properly speaking the ordinary is now in the midst of us. The things we love are not around us, but within us. We are physically separated from them, and we must now rely on the images granted by memory. We can see this process of the internalisation of the ordinary in the work of the Russian film director, Andrei Tarkovsky. Films such as *Solaris, The Mirror*, *Stalker* and *Nostalghia* are concerned with the notion of returning: of the need to return, of the possibility of getting back what we have lost, but of the *im*possibility of any physical return. Instead, what these films tell us of is the fecundity of memory and how it is located in special places, which are detailed and frequently obscure in their particularity, yet nevertheless poignant and evocative. Moreover, Tarkovsky, through the lingering on tiny detail and the slow, drawn-out nature of his storytelling, is concerned with the consequences of return, of what could and would occur if we were able to go back to what we have lost.

This notion of consequences plays heavily in *Stalker*, where the stalker continues to return to the Zone knowing the impact it continues to have on his family. The Zone is an area apparently contaminated by alien invasion and which it is now forbidden to enter. The area is full of dangers and traps, and can only be successfully traversed by experts known as stalkers. Yet within the Zone there is a room that will grant to anyone who enters their most desired wish. The stalker, despite periods in prison as a result of being caught attempting to enter the Zone, continues to lead others towards fulfilment and enlightenment, even though it appears to have led to a disabled child; the children of stalkers, we are told, are often born with some disability. The stalker, then, has plenty of reasons to avoid the Zone, and at the start of the film we see him arguing with his wife, who attempts to persuade him not to return. Yet he regards being in the Zone – a forbidden, dangerous and mystical place – as coming home. This is the only place where he feels at peace with himself and the world. Tarkovsky demonstrates this by showing these scenes in colour, whilst all the other scenes are in monochrome. Also outside the zone there is cold, with ice and snow all around, yet the zone is temperate and fecund. But it is not a place where he can stay; he is there on sufferance, because he knows the rules and respects them.

What is interesting about Tarkovsky's portrayal of the Zone is that we see little in the way of supernatural or frightening occurrences. We only know of the properties of the Zone through the stalker's somewhat gnomic pronouncements of its dangers. Apart from some derelict tanks and other discarded military equipment, we see little that is not normal, if now overgrown and derelict. But this only serves to heighten the mystery of the Zone: it is both ordinary and yet a source of fear. Tarkovsky creates a deep sense of the particularity of a space out of what are quite normal elements: it is what they mean to the characters that creates their distinctiveness. As Peter Green (1993) has argued, despite his metaphysical and spiritual concerns, Tarkovsky tends to minimise the use of effects in his films relying instead on mood and impression. In both of Tarkovsky's so-called science fiction films – *Solaris* and *Stalker* – there is very little science and even less fantasy. His films are grounded in, and depend on, the ordinariness of the world.

Yet, in all his films, Tarkovsky connects up the mystical and the mundane, the ordinary with the spiritual. Indeed, he looks for and finds the spiritual in the ordinary. He imbues the apparently ordinary surroundings with significance, often by lingering on details so as to plant in this vision a sense of doubt: why are we lingering on this, what does it mean? We are forced to look at ordinary things in a new way and to connect them up with what comes before or after. This emphasis on simple observation, such as the movements of underwater plants at the start of *Solaris*, or the everyday objects discarded in a pool in *Stalker*, help to locate us – the objects can be seen to represent the modern world – but also to emphasise the importance of the particular. We are attached not to the notion of place but to *a place*; we connect with a particular object and not all objects. What Tarkovsky shows so well is that the ordinary around us is not mundane, but has a significance created by its association with us, and importantly that significance, often unlike the object or the place, is transportable.

This, of course, can be a source of comfort, in that we can remember what we have had and where we have been, but it can also be a regret, especially if we are unable to return. Whilst the stalker can and has returned to the Zone, where he can find peace, this is not the case with other characters in Tarkovsky's films. In *Solaris* we see an attempt to recreate and to rebuild a relationship. Kris Kelvin has been sent to investigate the strange happenings on the space station orbiting Solaris, a planet covered in an apparently conscious sea. When he arrives he finds only two survivors, neither of whom is co-operative, but both refer obliquely to strange visitations. Kris is himself the victim of such a visitation, in the form of his deceased wife, Hari, who committed suicide ten years previously, an act that Kris feels responsible for. But this Hari is not an apparition, but a physical creature, albeit one with no memories of the past. This scenario of a visit from a past lover gives Tarkovsky the opportunity to consider the nature of faith, and the possibility of returning to a past state and thereby attempting to undo our past actions. After Kris's initial panic at finding a physical embodiment of his dead wife, he tries to re-establish the relationship with this new Hari and, much to the disgust of the two scientists on the space station, he announces he loves her. Yet as Hari learns more of her situation – as she becomes more 'human' – she realises the impossibility of her situation. She can only survive in the vicinity of Solaris, and so she allows one of the scientists to terminate her existence. The film then is about the need to hold on to things, and of the consequences of our actions. We cannot undo the mistakes of the past, because we cannot forget what led to these mistakes and what came after.

But we should not forget even though we do regret: Kris seeks to hold onto this new version of his wife, even though he knows it is impractical and irrational. What Tarkovsky appears to be suggesting is that we should treasure our memories and hold on to them. As Peter Green states: 'Although past time may be irrecoverable, it cannot be destroyed or vanish without trace. Time and memory merge, are two aspects of a single phenomenon' (1993, p. 69).

We see this in another important theme in the film. This is the way Tarkovsky contrasts the naturalism of Kris's home at the start of *Solaris*, with the plastic artificiality of the space station. Mark Le Fanu (1987) notes that the first third of *Solaris* takes place on earth at Kris's dacha and in his garden. The sense given by this old house and large, wild garden is to 'establish for the film a terrain of memory, a human landscape of loved things to be looked back on with longing and desire' (p. 54). This is an example of a chronotope, where place is located by specific time-bound sensations. We also see that many of the same artefacts – fine art prints, old books, busts of philosophers – are found in both the dacha and in the space station's library. Tarkovsky consciously tries to link his character to a tradition and culture, and in doing so show how we are linked to our own past. We cannot rely on science, or trust to the future, but rather we must store up artefacts from the past. Green concurs with Le Fanu about the start of *Solaris*, when he states it is:

> as if Chris Kelvin were trying to capture in his mind the essence of the planet and his life there before his voyage […] Kelvin fixes the co-ordinates of his life in his mind – the memories of childhood, the lakeside landscape of home. They are images to which he will return in his thoughts and dreams. He washes his hands in the water of the lake. It is a conscious leave-taking; for his voyage to Solaris is also a journey into exile from which he may not return (1993, p. 70).

When Kris leaves for a precarious journey into the unknown, he wants to take certain things with him. These things are lodged in memory and can become reference points. The re-embodied Hari is such a focal point, for she has been resurrected out of Kris's memories by the active brain-sea of Solaris.

The film, then, is not really science fiction: the scenario is merely a vehicle for a analysis of the importance of the past. As Green (1993) states, 'The film revolves for much of the time around the recollection and re-creation of terrestrial phenomena in an alien environment' (p. 72). Green suggests that Tarkovsky's point is that 'in rescuing the ideas and objects of his cultural identity, man rescues himself from oblivion' (p. 72). In the film's final shot we see Kris walk towards his father, fall to his knees and embrace his father's legs. The symbolism here is of the prodigal son's return, of the man of science returning to abase himself before his ancestors.

Perhaps the key conversation in the film is between Kris and Snaut, one of the remaining scientists on the space station. Snaut remarks that what humans are looking for in space is not other worlds, but themselves: humans are seeking their own image. They seek to go outwards to understand more of themselves inwardly. Green (1993) notes that in *Solaris*, as in several of his other films, Tarkovsky uses the situation of:

> removing his protagonists from their familiar, everyday surroundings to an alien ground – battlefield, forbidden zone, exile, or in this case, space station – where he proceeds to hold up a mirror to them and confront them with an unknown image of themselves (p. 65).

For Tarkovsky, the notion of exile is a device for examining what is meaningful in our spiritual existence. It allows us to note what is significant in our normal taken-for-granted lives by putting us in a situation where we can no longer be so sanguine. We are forced to look at ourselves, as if we were standing before a mirror. Accordingly, in addition to the eponymous film, this is a very common image in Tarkovsky's films. According to Green, it is used as 'the metaphorical looking-glass that provides man with a reflection of himself. In its surface, time is refracted; and it is a transitional device through which one may pass to other worlds, other states of consciousness' (p. 80).

What is shown through the looking glass in Tarkovsky's *The Mirror* is a particularly delicate portrayal of houses, with its attention to the small detail of the rural dwelling, with its exploring kittens, billowing lace curtains, spilt milk, falling objects and childish tricks. All of these experiences are so particular in their details as to be pure memory: they are not stereotypes but singular memories. This is because they are autobiographical, but as Le Fanu (1987) has stated, '*The Mirror*, despite being personal, speaks somehow with the authority of third-person narrative art. Autobiography in the film is woven into history, lending it a grandeur and a classicism' (p. 69). The film is episodic and has no distinct chronology, and Tarkovsky makes no attempt to tell his story directly. But it is precisely this impressionistic quality that is so evocative of memory. As Le Fanu (1987) has said, 'One of the central strengths of *The Mirror* lies in its simple power of evocation: its ability to conjure up, in piercing epiphanies, that magical submerged world of wonder which forms the adult's later imaginative capital' (p. 78).

So *The Mirror* is concerned with memories of childhood, of wishes unfulfilled and regrets. It shows the cyclical nature of the ordinary. Natasha Synessios (2001) sees much that is autobiographical in the film, in particular the inability to act as father to one's children. She suggests that neither Tarkovsky nor his father could sustain family life. They both left their first wives and children. Synessios states that, 'The home of family was not the one where they felt most at home, though Tarkovsky carried within himself a life-long nostalgia for an idealised family and home' (pp. 85-87). We might see that this idealisation derives from the double failure of father and son, and as a form of compensation for the loss and lack. Of particular interest in this regard is the use that Tarkovsky makes of his father's poems in the film. In one poem, *Life, Life*, we hear the phrase, 'Live in the house and it will not fall down'. The implication here is that if we believe in family and keep it strong, then we shall remain. This theme resonates with the same regard for the culture of our ancestors that we saw also in *Solaris*.

In this sense we can see the film being concerned with hope, faith and redemption. It is how we as adults look back to our childhood memories, with its sense of home and familial security. The film shows that 'Our past is our fortune. We exist as moral beings insofar as we possess, love and imitate ancestors' (Le Fanu, 1987, p. 73). The film goes through a series of scenes concerning the life of Andrei, whom we see as a small child before the Second World War, and as a teenager during the war. The central figures of these scenes are the boy's mother and the absent father. These scenes are interspersed with others set in the present day of Andrei's wife and teenage son, Ignat. In these scenes we only hear Andrei, but do not see him. The connection between the generations is made, and the circular sense of family heightened, by the use of the same actors for both teenage boy characters (Ignat Daniltsev) and the mother/wife figure (Margarita Terekhova). Some of the scenes are intended to represent actual happenings such as the burning down of a barn, whilst others are dreams and visions. The film tells no particular story, but can be seen as a collection of memories and wishes, fulfilled and unfulfilled. It is about an adult, full of regrets, who can no longer experience the innocent sense of possibility that comes with childhood; of a person who is *dis*located. Green (1993) suggests that:

> the view Tarkovsky seeks is that of the child, with which we glimpse Utopia or paradise. The point in man's history where he takes the wrong path is where the child loses its innocence and begins to comprehend the world in documentary form (p. 85).

We do not remember our lives in a linear manner, viewing one incident following another, but rather as a mix of the actual and the hoped for, of promise and regret. Thus Green sees that *The Mirror*:

> marks an attempt to recover the vision of childhood as well, not just the memories, but the unexplained mysteries, with all their discontinuities and distortions of time; a child's-eye view of the world and history, which accounts in part for the elusive fascination and haunting quality of the film (p. 85).

Memory can be a place of safety as well as yearning. We may have regrets and wish things were different. But whilst we might want the past to be changed, it can also be a source of comfort that things must remain as they are. We can enhance the past but it will always remain there for us. The virtue of memory is that it can be sealed up. In trauma or anxiety we can either dwell on the comforting or the regrettable. There are things we can cling on to which offer us security and then there are things which build the load.

The seminal scenes in *The Mirror* take place within particular dwellings, in particular the rural idyll of the pre-war family, and the apartment of the grown Andrei. The dacha was modelled on the place where Tarkovsky himself stayed as a young boy with his sister and mother. It offers a tremendous evocation of the warmth and simplicity of childhood. The scenes are full of detail, from spilt milk and tiny kittens to flowers in jars on the window sill. These scenes are largely shown in warm colours which show the varnished wood and intricacies of the lace curtains to the full. It is clear that these are memories of security and comfort. The child may feel the lack of his father even here, but this is balanced by the strength of maternal care and the fecundity of the natural world that surrounds the dacha. This is then a remembered idyll, the childhood paradise, in which there are few troubles and things need not be understood, only felt. Indeed, several commentators on the film have suggested that the film is best felt and not analysed: it is a film of impressions rather than arguments.

This childhood home can be contrasted with the apartment of the grown Andrei. We never see him, other than a fleeting glimpse of his arm on his death bed at the end of the film. However, we do see his apartment, in scenes with a group of Spanish émigrés, in dialogues between Andrei and his wife, and scenes involving Ignat. Just as the childhood dacha is cluttered and cosy, the apartment is large and well lit. In one scene we see no inhabitants in the apartment at all. Instead the camera pans through the apartment as the unseen Andrei talks to his mother on the telephone. She reminds him of one of her work colleagues who has recently died. He struggles to remember the woman, and it becomes clear how mother and son have grown apart. As they talk, barely connecting with each other, we are shown his apartment, and we see it as in some way tarnished, as lived in, yet also sparse and empty. It is impersonal and anonymous and has also been somehow reclaimed by nature: the walls are spalled and the wood is marked and worn. There is a history to this place, but there is still no real human trace. Rather there is a coldness, a sense that the dwelling is not fully occupied. Just as the film lacks the presence of the grown Andrei, so the apartment lacks any real human presence. The most pressing sense here is one of neglect, of things left to decay. This can be seen as a metaphor for the state of the human relationships within this space. In Andrei's dialogues with his wife it becomes clear that they are separated and that Andrei rarely sees his son. His only conversations with his son are on the phone and a brief question over whether Ignat wants to live with him rather than his mother. Ignat's response to this is one of alarm. So this place, whilst it is on occasions brightly lit, is somehow empty even when it is populated: there is a vacant space at the heart of the dwelling, just as there is a hole in the relationships between father and son, and husband and wife.

We thus see how a singular place can be used to evoke opposite sensations: on the one hand, we see a rural idyll built by memory; and, on the other, we sense a place in decline as the relationships fail. One is a place of possibility, whilst the other is one of regret, and they are linked in that the memory of childhood is necessary precisely because of the failure of the present. We need not see the memories as accurate as this need not be how the child's home actually was; memories are always partial. What matters is that we need something to cling to when we are alienated. We need to believe that if 'we can live in the house it will not fall down', and if we can keep it filled with memories this will serve to do just that.

The Mirror is a film about exile, about the effects of the distance between comfort and reality. It is concerned with what we have lost when we leave our childhood innocence behind and have to bear the complexities of the adult world. What it shows is that our sense of the ordinary does not just depend on where we are, but on what we have. This allows us some ability to adjust: when we are alienated from our immediate environment, we can draw on our memories, and use these for our dreams and visions.

This, of course, hugely complicates our view of the ordinary. By bringing in memories, and showing how evocative they can be, the ordinary becomes something both larger and more nebulous. This makes it harder for us to describe and to categorise how we experience the ordinary. But it does hugely enrich our view of it. The impressionistic quality of this particular Tarkovsky film demonstrates that the ordinary is a mixture of memories of places, things and people. The ordinary therefore encloses both the past and present, and we can see that one can compensate for the other.

The Mirror can be seen, and should be seen, as a claim for the resilience of the ordinary: it is a study of how we can use our resources of memory to sustain and justify us. Memory we can see as a form of compensation, where we substitute a loss in the present with the stores we have built up over our life. This sense of compensation allows us to see the notion of home as located within and bearing on whom and what we have. When we look at notions of home we have to include our longings, dreams and memories, as well as the concrete physicality of place. So, we should suggest that *The Mirror* shows the strength of the ordinary and the resources we can call upon.

Nostalghia, Tarkovsky's penultimate film, continues this consideration of compensation. However, it might be seen to present a more extreme case. Instead of a man being alienated from his past, the hero of *Nostalghia*, Gorchakov, is in physical exile, being in Italy away from his native Russia. The view in *Nostalgia* is a glance backwards from exile to a lost home (Green, 1993, p. 85). Green suggests that Tarkovsky's aim with *Nostalghia* was to 'make a film about Russian nostalgia, a state of mind he regarded as peculiar to his compatriots when removed from their native land' (1993, p. 107). And again, we are given a picture of place beyond the purely physical: 'home is also a place within the heart, a scrap of language, lines of verse that cannot be translated, memory, time past or visions future' (Green, 1993, p. 108).

Nostalghia was made by Tarkovsky in Italy, and is concerned with exile, just as the director himself now found himself exiled from Russia and separated from his son. The film centres on a Russian poet, Gorchakov, who is doing research on a Russian musician who spent some time in Italy before returning to life as a serf in Russia. Travelling with his Italian interpreter, Eugenia, Gorchakov arrives at a Tuscan spa, where he meets a local madman, Domenico, who sets him the challenge of taking a lighted candle across the village spa. We discover that years before, Domenico, fearing the end of the world, had locked himself and his family in their house for seven years. Gorchakov recognises that what Domenico has is not (just) madness, but faith, and is struck by this drive for redemption that he sees within him. Following his encounter with Domenico, and Eugenia's return to Rome, Gorchakov goes through a personal transformation, where we as observers struggle to separate dream sequences from apparent reality. But the result of this inner quest is that Gorchakov resolves to fulfil Domenico's request. In the climax of the film we see Domenico preaching to the disinterested citizens of Rome that they must repent and begin anew, before setting himself alight. These scenes are edited with a very slow and painstaking scene where we see Gorchakov walking with a lighted candle across the now drained spa. As he reaches the far side, he succumbs to the heart complaint that has been suggested earlier in the film.

Throughout the film Gorchakov dwells on his idyllic Russian home, a small rural house set on a hill, where his wife, mother and children await his return. These scenes are in sepia and in soft focus to represent memory and longing. In the very final scene we see Gorchakov sitting beside his dog in front of the remembered Russian house. But this whole scene is now within a ruined Italian Cathedral. In an earlier dream sequence we witnessed him walking through these ruins. Gorchakov, in death, has found home within the broken grandeur of faith.

The film is beautiful to watch, yet steeped in symbolism and imbued with mystical significance, and it resists any easy reading of its meaning. As with *The Mirror*, this film is not a conventional narrative, but is rather an inner journey in which Gorchakov finds a means of coming to terms with exile. It is often not clear whether we are watching reality or a dream. But, as with Tarkovsky's earlier films, we should not seek to be overly analytical, but instead wallow in a sea of impressions prepared by the director.

Nostalghia is concerned with the manner in which the memory of the exile idealises a dwelling and the ordinary. The exile lives in a netherworld disconnected from his now non-existent home and the place where he lives, which he cannot see as home. Gorchakov talks to Eugenia about the impossibility of translation, where meaning cannot be translated from one place to another. We are shown this vividly in a dream sequence where Gorchakov falls asleep in his Italian hotel room and remembers his house in Russia. We see both his wife and Eugenia in the dream. Yet his longing for home and family is so vivid as to wipe out any desire for his translator, and we see Eugenia comforting Gorchakov's wife instead of any consummation of his desire. Eugenia represents the seduction of Italy: 'the threat that he might forget his own roots' (Green, 1993, p. 110).

Gorchakov is unable, or unwilling, to settle in his Italian hotel room, or to get close to his new companion, despite her efforts. Instead he dreams of his wife and his remembered home. This scene comes after his refusal to visit a church Eugenia has driven him to at his request. Gorchakov's memories are all of a particular place, shown in sepia: this is a defining space where he longs to be, and which he only achieves in death at the film's conclusion. The film can be seen as being about loss of place, but also an inability to settle because of what he cannot return to, yet which he cannot forget. What he yearns for has no intrinsic beauty, certainly not in comparison to the sights of Italy we are shown. The sense of beauty is within him – he gives it this sense – because of what it encloses, what it means and because of who is, or was, there. Le Fanu (1987) notes that Tarkovsky refrains from portraying Italy as a place of sunlight and lemon trees: 'The film's gloominess comes from a kind of resistance to Italy' (p. 112). This is because the film is concerned with the question, 'can an artist survive and flourish in a country other than his own?' (p. 112). For Tarkovsky seems to be saying that 'the real place of the heart, the place from which alone the artist draws sustenance and remains alive, is bound to be his own native land' (p. 112). It is significant that Gorchakov's reason for being in Italy is to research a musician who returned to Russia from exile even though he knew he would return a serf. This was because he too yearned for a woman he loved and for the place in which he felt was his home.

Tarkovsky portrays nostalgia as a sickness, which gnaws at the body. Throughout the film Gorchakov's heart condition is referred to, and his death at the end of the film has huge significance for Tarkovsky's understanding of the longing for home. As Green (1993) suggests, Gorchakov, 'heart-sick – or sick at heart – dies of his illness far from home: nostalgia as sickness for another place, another time, another state, so severe as to amount to a disease – a sickness unto death' (p. 111).

What we see in *Nostalghia* is that the ordinary is made special by distance, when it has to be transported into the present by memory. Distance and separation make the ordinary different and unique. This is because it is, properly speaking, no longer what we are (physically) in the midst of. Instead of fitting in with our immediate surroundings, it now stands out and jars or grates against what we are in. But the effect of this is to diminish the present, not the past. It is Italy that is made gloomy, wet and covered in mist: it cannot compete with the idealisation of home drawn from memory.

Can we see the problem of the exile as an inability to accept new surroundings as ordinary? The exile cannot leave behind one sense of the ordinary and then reduce the current environment to normality. Instead Gorchakov nostalgically builds up the old sense of ordinary into something special: as an exile he clings on to an idealised sense of what he had, and almost reifies and then deifies that sense of the ordinary. The nostalgic person cannot therefore integrate the new surroundings into themselves, but tries to impose the old sense of order on to the new environment. This is why Tarkovsky sees it as a sickness, because there can be no cure without the possibility of a return.

The role of Domenico, the apparently mad loner, is important in this story of loss of home. We hear that, fearing the apocalypse, he kept his family in his house for seven years. He sought to protect them from the world but, of course, all he did was to entrap them and to circumscribe their experience. Here we have the opposite to Gorchakov's yearning: Gorchakov seeks enclosure and a return to his womblike home (with its sepia tint and amniotic like mist), but Domenico's place is a prison, even if it is one born out of familial love and a desperate need to care.

And as we see the ruins this made of Domenico's life, we also see the ruins of his dwelling. His wife and children left him when they were released and he spent time locked away in an asylum before returning to the family home alone. The house is now derelict and full of water. Domenico's house only maintains vestiges of habitation; a door to the main room remains, but there are no walls. But whilst Gorchakov steps over the ruin of an internal wall to move from one room to another, Domenico pointedly uses the door, opening and closing it with deliberation and love of place. This is still his home and as a store of memory it deserves respect. This is still the centre of Domenico's life, and Tarkovsky shows it to us with a tremendous attention to detail in terms of the *mise en scène* and the sounds we hear. The house, partially submerged and ruined, tells a story of regret and loss. Green (1993) suggests that water denotes not just the traditional concept of purification, but is also associated by Tarkovsky with the idea of home and homesickness. Hence there is a pond before the Russian farmhouse, and the dream sequence takes place whilst Gorchakov is beside water within a ruined building. Similar associations are made in *Stalker* where we see the stalker himself sleeping on a tiny island amidst a pool of water. Also in *Solaris* we see Kris Kelvin lingering beside a pool in the garden of his dacha before setting off on his long and perilous journey into space.

Domenico's role in the narrative is as Gorchakov's alter ego, the alternate side of his character that represents faith and the search for redemption, instead of longing and nostalgia. We are shown this in one of the dream sequences when Gorchakov looks in a wardrobe mirror and sees Domenico instead of himself. According to Green (1993), 'If Gorchakov's nostalgia is a deadly sickness, Domenico's is a yearning for life, for a better world that transcends death' (p. 114). And it is through Domenico that the Russian finds peace in his exile, and we see a reconciliation of Russia and Italy. We see this at the end of the film, in what Green (1993) sees as the realisation of 'a landscape of the mind' (p. 114), where Gorchakov sits with the dog in front of the house of memory and within the ruined cathedral. He has arrived at 'the place where one's inmost wishes are fulfilled' (p. 114). What Gorchakov gains, through Domenico's example of faith is a form of *acceptance*. He finds somewhere to place his memory and hence we see him in front of the dream house but within the ruined Italian church. The two elements now can come together and we see what he finds faith in.

What Gorchakov attains, and what is also sought in *Solaris* and *The Mirror*, is the stopping of time. Tonino Guerra (2004), the co-writer of *Nostalghia*, describes an encounter between the Italian film director Michelangelo Antonioni and some elderly women in Uzbekistan:

> Antonioni, too, made great use of a Polaroid at the time, and I remember that during a reconnaissance in Uzbekistan for a film that in the end we never made, he wanted to give three elderly Muslim women a photograph he had taken of them. The eldest, after casting a brief glance at the image, gave it back to him, saying, 'Why stop time?' We were left gaping in wonder, speechless at this extraordinary refusal.
>
> Tarkovsky often reflected on the way that time flies and this is precisely what he wanted: to stop it ... (p. 7).

This is what it means to remember and to search the memory: we fix something in time, to stop it progressing to where we might not like. What we want is to encapsulate something in its literal sense, of putting it in a capsule so it cannot be touched, so it is protected and preserved. In this way, we have captured it in its wholeness and it can progress no more. We remember fixed points just as a camera can create them. This is the way in which Tarkovsky captures places and their significance to us. He seeks to stop time and to hold onto those precious artefacts created within time, and which become solely ours once they are held fast in memory. They are now linked intrinsically to a specific location (Bakhtin, 1981). In this way Tarkovsky provides a profoundly robust view of the ordinary. The ordinary becomes transportable, and thus we can take our common place with us. We can overlay the alien world with our ruts, just as we see the track leading to Gorchakov's idealised Russian farmhouse, and we see this house rooted in an alien soil. In this way the alien environment can become home.

No Place Like Home

One of the problems of so-called 'popular entertainment', be it Hollywood or elsewhere, is that it must be action-packed. The story must develop with a pace and things of consequence must happen. The film should have a clear beginning which sets up the story and gives us the backgrounds of the characters (all of which, of course, is later significant); it should have a middle in which the story develops into something decisive; and then the end in which all is resolved and explained, and where the problem is transcended and the right sort of values come out on top. The essence of the popular movie, therefore, is one of progression, of things moving forward towards some resolution.

Yet the ordinary is not fast-moving. Instead we would see it rather as a static relation between us and our dwelling environment. We do not want to move, or to change, or to have things resolved. There is no need for this, as things are already as we would like. And, in any case, we seek no outside interference, no big upheaval that forces us from our current position to another. We seek to move when we wish to, and stay where we want. Despite the facility of memory, we would rather live amongst the people and things we love rather than merely having to remember them: we wish to dwell *in* and not just dwell on.

In contrast to the movies, our sense of the ordinary has no beginning, except in the obvious sense of our birth. Nor can we see it ending, even though we know it will. As Stanley Rosen (2002) has suggested, the ordinary is all middle, in that we are always in the midst of something. There may be peaks and troughs, happy occasions and sad, but there is no end: as Epicurus nearly said, where we are, the end is not.

Part of what we want in the ordinary is *stillness*, a sense in which we are not put out and where we are not contested by the threat of transformation. We seek tranquillity and stability in our private dwelling as a bulwark against the fast-moving and bewildering world outside (King, 2004b). This stillness is a consequence of the ordinary's comprehensiveness and its unity. The ordinary is of a piece, where no parts of it protrude, or move at perceptibly different rhythms. One part does not appear to change any more than the other parts or as it does as a whole.

Moving images can also exhibit this stillness. Certain films, a few of which we have already considered, contain a lack of resolution and a slowness that seems to match this still sense of the ordinary. The films of Andrei Tarkovsky move slowly and linger on details, and this process memorialises the banal and apparently insignificant. But we can also see it in the work of other film makers, such as the Hungarian director Bela Tarr, whose films move slowly, showing a fascination with the faces of ordinary people and the rhythms of lonely individuals moving through their decaying urban environment. Like Tarkovsky, Tarr shows the inner condition of his subjects through the environment around him. His films, such as *Damnation* (1987) and *Werckmeister Harmonies* (2000), show a bleak world of unhappy people bemused by change and who fear being left behind. They seek some security in relationships or in places, but find little comfort. And all around them is a bleak, cold and damp environment that appears to be sucking the life out of them. This environment, particularly in *Damnation*, creates a sense of inertia, in which the film's main character, the listless and dissolute Karrer, pursues his obsession with a nightclub singer, whose ambitions go beyond the small town she is currently stuck in.

Another example of this form of stillness in cinema can be seen in two films, *Gerry* (2002) and *Elephant* (2003) by the American director Gus Van Sant. Van Sant readily admits to being influenced by both Tarkovsky and Tarr, and these two films both have the same slowness and refusal to resolve situations. What Van Sant's two films show, as do those of Tarr, is that there can be no ready resolution in our ordinary lives, and that there is unlikely to be an ending that unites all the strands in our own particular narrative.

But the two films of Van Sant are useful in a further sense. Not only do they show how the ordinary is all middle, but they also point to the limits of the ordinary. In *Elephant*, we are shown what occurs in the ten minutes preceding a shooting in a school similar to Columbine high school massacre. We follow a number of school students through the bleak, antiseptic corridors of a school: the camera follows one individual or group, and then cuts to another situation involving another individual or group. These strands come together briefly and then move apart again, until we see the two gun-toting students enter the school and start shooting staff and students in the library. We are shown occasional flashbacks, particularly of the two assailants, but there is no attempt to explain or come to a judgement on them. We are no nearer understanding their motives at the end of the film, nor do we see how the situation is resolved.

The violence at the end of *Elephant* has a much greater impact because of the first hour with its slowness and repetition, which serves to emphasise the ordinariness of the characters and their situation. What creates the sense of foreboding is the very triviality of the encounters, their banality and lack of obvious action. As would have been the case in Columbine, none of the victims has any inkling of their impending death, and so they act as on any other school day. Accordingly, Van Sant makes no attempt to place any significance on the actions of the characters, or to sow seeds of significance for later. He does not seek to understand the actions, but to locate them within the tiny details of the lives of the victims and perpetrators. Despite our foreboding and knowledge of the ending, properly speaking, no one but the perpetrators is so aware, and so they fill their day with their normal activities. It is the attempt to show these trivia, which are burst apart by mindless violence, that makes *Elephant* such a truly significant film. We only have to compare it with the absurd attempts to build drama and significance in murder mysteries, or the 'who shot JR?' saga of the 1980s TV series *Dallas*. What Van Sant shows is that all actions, even the traumatic ones, come out of the ordinary.

But this notion of being 'out of the ordinary' is one of degree: we can be mildly put out at times, but at other times we are so dislocated that we can no longer find the ordinary we have come out of. There might well be some places that are so out of the ordinary as to inhibit the invasion of memory: somewhere that is no place like home. This situation is portrayed in Van Sant's film, *Gerry*. This film concerns two young men, both apparently called Gerry, who are travelling in their car along a desert road. They pull off apparently so one Gerry can show some landmark to the other. After a while they give up the attempt to find the landmark and turn back. However, they soon find they are lost in the anonymous scrub of the desert. The rest of film involves their trekking across various forms of desert, bare rock and salt flats looking for some form of civilisation. Eventually one of the Gerrys (played by Casey Affleck) declares he has had enough, and this causes the other (played by Matt Damon) apparently to strangle him. Soon after, the surviving Gerry sees a road and is saved by a passing motorist.

What is remarkable about this film, ostensibly a survival story of man against nature, is its emptiness: very little happens in the film, other than the men's slow disintegration. The film has a slowness and lack of incident, as well as a lack of dialogue. It conspires to create an emptiness similar to the wilderness the two Gerrys find themselves in. What works is the cumulative effect of repetition to build foreboding and a sense of disaster. Many of the scenes in the film consist merely of the two men walking in front of spectacular scenery, which they ignore in their distress. We are also offered many shots of the wilderness and the sky using time lapse photography to show the effect of time passing (a technique also used in *Elephant*). But it is this very sameness of the scenes that is effective, in that it shows how an apparent sameness can be differentiated by the changing context. We shift from whimsy and the solution of problems, for example when one of the men is stranded on top of a large rock, to complete physical and moral degradation, shown by their physical collapse and the murder.

One scene in particular demonstrates this slowness quite brilliantly: all we see for nearly four minutes is a closely cropped shot of the bobbing heads of the two Gerrys as they trudge across a barren landscape, and all we hear is the crunching of their boots on gravel and their breathing. Their faces are sunburnt, they are tired and drawn and have long since lost any sense of camaraderie. There is no contact between them, no attempt to support each other. They walk in a strange rhythm which emphasises this separation, sometimes they are in time, their heads bobbing together, at other times one has moved slightly ahead, and then the other. We have no sense of a race or that either has any consciousness of the other beside him. Rather they just trudge relentlessly across an equally relentless landscape. What they are doing has a purpose, but has no end for them. What we see in this incredibly simple shot, with no dialogue and no development, is the loss of hope: they are moving forward but, because the camera is keeping pace with them, they are going nowhere. There is no heroism, no great escape, just human stupidity, stubbornness and perseverance. What we have are ordinary vices and virtues in an uncommon setting, and the juxtaposition is both engrossing and terrifying. So even in this barren, hostile wilderness all that exists are the ordinary human qualities, and there are no possibilities beyond this.

This scene is similar to one in Bela Tarr's *Werckmeister Harmonies*, where we see just the heads of two characters walking along a road for several minutes. These characters are on an errand they neither sought nor can avoid, and we see their dread and doubt along with their resolution to complete the task. They are not heroes either, nor do they exhibit anything but human frailty in the face of forces greater than them.

What Tarr and Van Sant show is that there is no ready distinction between the exceptional and the ordinary. In *Elephant*, the children in the high school did not know what was to come and so it would be crass of the director to build up the portents of doom. Extraordinary occurrences come within ordinary days, which are no different from others until the extraordinary occurs. There is then no build up, no ability to shift significance from after to before. Tragedy grows out of the ordinary just like all other occurrences, and we should not read too much into the ordinary. Likewise, the predicament of the Gerrys has arisen out of their carelessness and has no significance other than that. They began their trek with confidence and they stride across the landscape with evident purpose. They are sure of where they are going, even in an environment with few markers and which is unknown to them. They are playful and boisterous. Yet this is the main thing that fades from them as they proceed into the wilderness. They can no longer take the environment for granted and their certainties disappear along with their bonhomie, humour and communication. Instead they stumble onwards without any apparent purpose other than to get out.

There is an implacability to them in this hostile environment. There is no place for them here, there are no linkages with their ordinary lives, and no ability to derive sustenance from memory. They have no means with which to enter the ordinary. They can put down no roots, and they can find no ruts. There is just formlessness and a lack, which leaves them without any ordinary means to survive.

The significance of Van Sant's portrayal is that it shows the limit condition of the ordinary. It shows that there are certain conditions that make an ordinary life impossible, and unlike Tarkovsky's exiles, we cannot draw on our store of memory as a prop. We, of course, have nothing other than our ordinary experiences and capabilities, but these may not be enough for us. Our ordinary actions need their place and so the formlessness – the *dislocation* – we see in Van Sant's *Gerry* is unsustainable for us. If we are so much out of place, as in this film, the ordinary sense which we may carry within us is no longer of any use to us.

But the real tragedy for people in this situation is that they have nothing but the ordinary to support them. The two Gerrys do the normal things one would expect. They try to walk out of the wilderness and to civilisation. They take their bearing from the sun and seek shelter when they can. They use high ground to reconnoitre the landscape. But there is no heroism, and no transformation: the two men do not act differently from the way their natures dictate. Adversity is not the making of them; they find no hidden powers of resourcefulness. Instead, the situations in front of them are new to them and remain so. They are unprepared, physically and mentally, and so what we see is not heroism but mental and physical decline in the face of an extremely hostile environment. Despite being in this perilous position, they have nothing to rely on but what would ordinarily be at their disposal, as if they have taken a walk in the park.

We can compare the two 'Gerry' characters in Van Sant's film, with other 'survival' films such as John McTiernan's *Predator* (1987), where its lead character, known as Dutch (played by Arnold Schwarzenegger), survives against an apparently invincible foe, whilst all others fail. But fortunately Dutch is an elite soldier and so is able to use his apparently unique abilities, great strength, ingenuity and unbelievably good luck to defeat his enemy. In this sense, this movie is typical of the form we discussed above, with its need for the 'right sort' of resolution. Yet in *Elephant* there is no escape, no re-balancing of the odds, and individuals cannot go beyond the limitations of their ordinary selves. Despite the tenacity of the victims, such as Chris, who we see stalking one of the killers and who we think may be on the verge of overpowering him, they too fail and suffer. The killer sees Chris and shoots him. Ordinary people, who are untrained and with no awareness of fighting, weapons and conflict, are no match for those prepared and with powerful automatic weapons. We cannot fight against such odds. Van Sant shows us that the ordinary is a place of limits, and we cannot go beyond these limits, no matter what the stresses we are put under. We might wish that we can face up to adversity and that it will bring out as yet unknown faculties within us. But we fear, or even know, that we have no such faculties and that, faced with an extreme threat, we would have nothing to rely on but our prayers.

The key message in Van Sant's films is about our inability to deal with transgression, to cope with changes imposed upon us. We need to remind ourselves that what Deleuze and Guattari (1988) described as transgressive acts – the results of so-called schizoid desire – are really impositions on others, and that through acts of transgression the perpetrators impinge on others, imposing their wills on them, which takes the victims out of their sense of the ordinary. The two killers in *Elephant* seize the moment and create a singular event, but this, of course, is murderous to others. One person's transgression is another's death, or life ruined and certainties shattered. And for whose benefit? It is surely only for that of the transgressor. As I have discussed in *Private Dwelling* (King, 2004b), transgression is all too often merely a question of power, and therefore the very opposite of what is intended by post-structuralist theory. Transgression is an act of power and involves the displacement of order. But the ordinary depends on order, on definitive roots and clear ruts leading us from place to place. We do not wish to have roots pulled up and our traces scrubbed off the face of the earth, even if we are told it is for our own good.

The ordinary has its limits, and for most of us, for most of the time, we are safely within those limits. Even when we are in exile we can draw down from our store of memory and seek to rebuild our sense of place on what might be alien terrain. This is what successive generations of villagers in Akyazi have been able to do. They have made something permanent out of what is now local and what is still remembered. Less positively, we have seen how the characters in Andrei

Tarkovsky's films have sought to deal with their temporary and permanent exile, be it from the lost places of childhood, their country and culture, or earth itself. In all these cases, memory can help to deal with the alienation of exile. And more extreme still, we have seen how there are places where the ordinary can be of no help. Yet even here we have seen that we have nothing else but those ordinary faculties to sustain us. Unfortunately, as we saw when looking at the films of Gus Van Sant, there are times when this is just not enough.

My aim here is not to imply any sense of environmental determinism. I am not suggesting that there is only one place where we can thrive and survive (even if Tarkovsky could be accused of coming close to this at times). What I wish to suggest, however, is that there are limits to the tensile strength of roots; and that if we have no ruts, no tracks to follow, and if we do not know the way, then we will get lost, and perhaps stay lost. Sometimes we may have a map and can improvise from what we carry with us, but there are limits to what we can carry and how we can respond to extreme adversity.

But this does not detract from the ordinary – after all, some of our most treasured things are fragile and easily broken. But rather it further enhances the picture of the ordinary that has already been built up. If we are bereft of place and our sense of roots and ruts, we are nothing but mere shuffling shells of humanity and we have no sympathy with, or from, our environment. This, I would seek to argue, shows the importance of the ordinary: not to have it is to have no place. Without the ordinary we can only seek to survive and undertake the most elemental of activities. There is no place where we can stay and put down any roots. Instead we are forced to keep moving and to strive to get out of where we are.

We have a desperate need to find comfort in these places, but we have no routes to find our way out: we have no certainty as we are homeless and alienated in a desert. These surroundings may be beautiful and varied. But we can only appreciate this when we have the proper equipment, when we have transport, food, water and shelter. We can also appreciate these scenes whilst watching in a cinema (and without the proper equipment). But if we were out there, the beauty would be lost. So we should return to those places where we can find comfort, and these are the ordinary and ordered places we know so well. It is where we are likely to watch these films, safe in our own homes, in front of the home cinema system, watching with those we love and care for. It is where we are comfortably within the limits of the ordinary and free from the extremities of exile, and where we can see fictions for what they are.

Chapter 6

Accommodating Change

What allows us to enjoy, or at least bear, films about exile and alienation like *Solaris* and *Gerry* is that we can watch them at our leisure, be it in the cinema or at home. Were we with Kris Kelvin orbiting above Solaris, or in the desert with the two Gerrys, we might not be able to appreciate the nuances of the situation in quite the same way. It is precisely because of our current comforts that we can understand the problems of alienation and exile as presented by Tarkovsky and Van Sant: we have the leisure to reflect on these artistic attempts to portray the loss of home and the inability to find a means back to it.

Whilst we would want to agree with Tarkovsky on the indispensability of art and culture, this observation does tell us something about the nature of dwelling: what gives us the general ability to reflect on a trauma is often the very absence of its cause. We can take for granted our dwelling just as it supports us and allows us to enjoy activities such as watching films and reflecting on their significance. There is a lack of imperative in this state, where housing is our background, and so we can foreground other things that engage us.

But what if we seek to place our dwelling in the foreground and see it as the essence of our being? This is where instead of seeing housing as ordinary, we seek to create the extraordinary, to create something special, distinctive, and to transform our dwelling fundamentally. The problem with things in the background is that we might not notice them: taking things for granted is not always a positive, but might be a case of forgetting.

I want to suggest that housing is particularly prone to this act of forgetting: that as we sit in our comfortable dwellings, entertaining or educating ourselves, we forget what provides the comfort and holds us in this place. This act of forgetting is not an omission, but an action. By that I mean that, in forgetting, we do not merely fail to accept what we have, but instead we seek to replace it. It is precisely the problem that we do not or cannot accept the ordinary, and feel we must create something not merely out of the ordinary, but *extraordinary*.

The problem with our common place is that it fails to excite us: we see it as boring, as usual and it does not inspire us. It is just there and, in forgetting that is precisely what it is meant to be, we seek to make it into something other. What we wish to do is to 'makeover' our dwelling, to design it into something new. We feel that doing this will make the dwelling distinctive and stop us and, importantly, stop others from taking it for granted. In this chapter, then, I wish to consider the notion of design and the current penchant in the popular media for interior design or the domestic make-over. There is, of course, a considerable literature on the notion of design, much of which, such as the work of Rem Koolhaas *et al* (1998) and Bruce

Mau (2000), shows a considerable ambivalence towards it. There is a sense that, even as we are suspicious of the commodification that comes with design, we are also entranced by the possibilities of creation and re-creation that comes with this cult of design (Foster, 2002). I wish, so to speak, to domesticate this ambivalence by investigating what might be called the epistemic conditions of design, and thus to question what might motivate this apparent desire to 'makeover' the dwelling, and what are the implications of this desire. This bears directly on our sense of the ordinary, in that it calls into question its ubiquity – why do we want to make something extraordinary if we are in the midst of the ordinary? If the ordinary is so important why would we want to remake it?

In answering these questions I want to suggest that we are more able to accommodate change than we might consider, and that whilst we might desire change, much of what we do is actually to mitigate its effects. I want to show this by a discussion of the effects of technology in the domestic dwelling. Despite the arguments that technology is changing the way we think and live, I wish to suggest that we readily assimilate change, and do so in a manner that does not fundamentally transform what we have. I wish therefore to posit what might be seen as a Burkean view of change, whereby we accept change where it is necessary to preserve what we have (Burke, 1999). In our personal lives we seek stability even when, or especially when, we are seeking to achieve our ends and create change. Along with Edmund Burke, we can see change as being an act or preservation, whereby we alter things only because it enhances those institutions and things around us which make our lives meaningful. We need some solid ground to stand on, or something solid to push off from. We need this solidity to give our lives shape. It may be that we are unhappy with our lives and want to change parts of it, but we still need some sense of stability, if only to react against. But even when we seek change, we do not want it to be permanent, and certainly not change for its own sake. We seek change for a purpose that will enhance us as persons, to better live our lives, and then we want the change to stop so that we can then enjoy what we have achieved. We do not therefore reject technology, nor do we allow it to dominate us: rather we use it to enhance the lives we wish to lead.

The issue, then, is again one of acceptance, of being able to come to terms with where we are and what we have. We may well feel ambivalent about what we have: we take the comforts it brings, yet we might yearn for something different, new and perhaps more exciting. It is where we might seek to escape from what we find boring, yet without actually leaving behind what we have, just like the children in *Spring and Port Wine*: they want to leave, but are all the time aware that they have a place to stay and that it will remain there and ready for them. They wish to leave home, but do not appear to consider where they will go to, and so, quite naturally, they stay and are happy for it.

This chapter is therefore the reverse of the previous one. In chapter five we considered attempts to create or maintain the ordinary when in exile or lost. We now turn to the situation where we are in the midst of the ordinary, but wish for something else, yet we seek to do so without losing what we already have. The image is therefore not of the exile or the lone traveller, but of the family on the sofa watching *Changing Rooms* and dreaming, at least for as long as the programme lasts, of a different type of dwelling.

Making the Extraordinary

According to Hal Foster (2002) everything is now designed, be it 'your home or your business, your sagging face (designer surgery) or your lagging personality (designer drugs), your historical memory (designer museums) or your DNA future (designer children)' (p. 18). Foster speculates that this notion of design might be the 'unintended offspring of the "constructed subject" so vaunted in postmodern culture' (p. 18). If, as postmodernists hold, we are made and remade by our social context, then why should we not open ourselves up to the possibility of design? If nothing is natural, why not seek to make ourselves and our surroundings in the most congenial image?

But, of course, just how new is this penchant for design? We might, along with Foster, seek to link it to postmodernism, as indeed other critics such as David Harvey (1989) and Frederic Jameson (1991) have done, but the notion of the designed dwelling goes back several centuries. Perhaps what Foster is referring to is the ubiquity of design, or its central status in modern capitalist economies. And, of course, unlike the situation centuries ago, we can now all join in. He may have a point here, but I want to suggest that the dominance of design is part of a more general attitude towards the dwelling (and one that can be as typical of modernism as the postmodern).

The house is essentially a boring, dull structure. It is a simple box shape that can be continually replicated without losing any of its functional capabilities. This, of course, is one of its virtues, but it also presents architects and designers with a challenge to their creativity. Accordingly, they seek to extend the possibilities of this boring structure and create something distinctive (Birchall, 1988). Likewise, those who live in the dwelling wish to make it into a place that says something particular about them. The desire for design can be seen then as the attempt to create a sense of place, rather than finding it. It is based on the premise that we can design a place that is personal yet distinctive. But in order to do so we find we rely on experts and external advice to create something which is for our personal, private use. This means there is a cost, which is that the dwelling becomes uniform, being based on the ideas of others. The dwelling stops, therefore, being ordinary in the sense of being one's customary world, and becomes a performance instead of a tool.

We can see this situation (and show that the problem is not a new one) by looking at Jacques Tati's film *Mon Oncle* released in 1958. The film centres on Tati's most famous character, M. Hulot, who is the uncle of the title. His nephew, Gérard, is devoted to him, but his parents, the Arpels, disapprove of the rather old-fashioned and bumbling manner in which Hulot comports himself. The Arpels see themselves as trend-setters and reside in a modern, labour-saving house in the newly built area of town. We see this newly built area encroaching on the old quarter of the town where Hulot lives in a small apartment atop of a rickety house, which he can only reach through a complicated series of stairs and corridors.

Hulot's dwelling is therefore the very opposite of the designed and micro-managed dwelling of the Arpels. Indeed Hulot's building has an organic quality to it, as if it has spread haphazardly as a tree would gain branches. There is no pattern other than makeshift utility. Just as the new quarter is replacing the old, so the Arpels see it as their burden to bring Hulot into the modern world by finding him a job and a wife.

The Arpel's house is built in the International style, with flat roof, clean horizontal lines, and large expanses of glass. The interior is open-plan, or 'all-communicating' as Mme. Arpel describes it; the furniture is minimalist and looks distinctly uncomfortable. The garden is manicured, with stepping stones between seating areas to ensure no one steps on either gravel or grass. In the middle of the garden is a pool with a water feature in the shape of a fish spouting water into the air. The garden is surrounded by a high metal fence with a gate opened only from the house.

It is significant to Tati's vision of modern design that the predominant colour of the Arpel's house is grey, in contrast to the colours in the old quarter which are vibrant and warm. All the houses we see in the new quarter are grey, as are M. Arpel's factory (which makes plastic), his car and his suit. We see the Arpel's lives as being bound by sameness and regimentation. The house is extremely ordered, and our first sight of Mme. Arpel is to see her cleaning any surface she comes into contact with, including her husband's car as he drives away. Likewise, Gérard is continually exhorted to be tidy and to put his clothes and shoes away. His life is regimented and controlled by his parents, who fail to communicate with him properly.

This notion of communication is important in the film, particularly between the Arpels and their son, showing the difference in the manner in which the boy is able to be free with his uncle. This contrast is as wide as that between the colours in the old and new parts of town. In the middle of the film the Arpels hold a tea party in their garden with the idea of matching Hulot with their extremely affected neighbour who has told them she is often alone. Mme. Arpel takes her guests around the house, showing off the modern and labour-saving features in the house. One of the guests, doubtless vocalising what the audience is thinking, states that the house is empty, referring presumably to the minimal and bare (but decidedly fashionable) décor. Mme. Arpel's response to this indiscreet comment is to say, 'No, it's modern … all-communicating'. This sense of the house being joined up, of all the rooms connecting is something of a mantra for Mme. Arpel, who states this whenever we see her showing people around: for her, it is the house's most outstanding feature.

The idea is that the house is planned and rational, and allows for interconnections between the various parts of the dwelling. Yet there are no relationships within the house: its design and formality inhibits communication. It is too antiseptic and deliberately kept so: everything has its place and should not be moved. When Hulot is seen sleeping on a sofa, which he has turned on its side to make it comfortable, M. Arpel considers it intolerable and an abuse of his property. But, of course, this is the only way the piece of furniture could be made useful and thus relate to the human occupants of the dwelling. The house, instead of being 'all-communicating' is empty and lifeless. Communication is precisely what is lacking in the dwelling, especially between Gérard and his parents. This becomes apparent when the boy goes out with Hulot. He is indulged and allowed to play with the gang of boys from the old quarter and get into scrapes with Hulot's friends. But once he returns home Gérard is criticised for his lateness and the state of his clothes. There is no rapport and no sympathy from his parents and this mood is enhanced by the clean, empty surfaces and minimalist design. The garden fountain in the shape of a fish, which spouts coloured water, is only turned on when a visitor calls and only then for valued guests. It is quickly turned off when tradesmen or Hulot arrive. It exists entirely to impress but, of course, only when the person is worth impressing.

What *Mon Oncle* demonstrates is that the designed property has a tendency towards rigidity. Once it has been created it should remain in its original state. In consequence it can show no signs of habitation. This is because the house is designed to be shown and not for comfort: the humans who reside in and visit the house must conform to it. We see this particularly with regard to the designed garden. The Arpels and their guests remain on the stepping stones and off the grass and gravel. This makes it difficult to walk side by side and so companionship turns into farce. When the guests are sitting in the garden the formality of the place is seen in the rigidity of the relations, with forced conversation and discussion of work. There is nothing that is spontaneous until Hulot arrives and creates chaos with his well-meaning bumbling.

The Arpels attempt to order their environment, and to impose a particular mode of living. But Hulot subverts this, not through any deliberate attempt or intent, but by simply acting in a way that does not appreciate or cannot accept the extraordinary: his main response to what is happening around him is one of surprise and bemusement. Tati's vision is not malicious: he does not seek to be sarcastic or to tear down his targets. Rather his view is to mock gently the conceits of his fellows. Hulot does not seek to change the world, but merely to remain within it as he is. He seeks his private bubble where he can be inconsequential, if rather bumbling. Problems occur when he tries to enter the worlds of others, be it a department store, restaurant, office or his sister's house. This is because he cannot adjust to the rigid patterns of modernity. The contrast could not be greater with the old quarter with its ramshackle, disordered, yet homely nature. The area which includes Hulot's house has no pattern or order, yet it has much more interest and life to it. It has some animation, informality and wit. We see this as the inhabitants of the house, and other locals, visit the local café, one man in his pyjamas and dressing-gown, and Hulot with his freshly laundered shirt under his arm. This house, and the environment more generally, accepts its inhabitants, whilst the Arpels' property does not. There is a conviviality to it rather than the formality of the Arpels' house, where casual contact is difficult and any engagement with others must be planned.

What *Mon Oncle* shows is the difference between creation and evolution, between making an environment and finding it. The Arpels' house is designed according to a particular vision – it is 'all-communicating' – whereas Hulot's residence gives the impression of having evolved over time and has not been planned or purpose built. The relevance of this film is that it sheds light on the nature of domestic design and the effects this has on the way in which we are presumed to use the dwelling. Tati contrasts homeliness with sophistication, animation with the antiseptic, and minimalism with vibrancy, both of life and colour. This points to the problem with design and the fetish for interior decoration that in the first decade of the 21st century seems to dominate the British television schedules.

This sense of the fetishistic is made by Marjorie Garber in her book, *Sex and Real Estate* (2000). She sees that the concern for houses and their interiors has replaced sex as the main preoccupation of the middle classes. This is because she sees the same sense of desire for property and furnishing as for the bodies of those we love and desire. The subtitle of the book is *Why We Love Houses* and she indulges in some loving descriptions of kitchen worktops and soft furnishing. Garber suggests that we have affairs with our dwellings, falling in love with them and carrying on the affair with an intensity, before having our eyes turned by something else that comes along. She is somewhat ambivalent about this herself: on the one hand, she is an academic and able to look at the phenomenon with a degree of detachment; but on the other hand, she too admits to being sucked into this circuit of desire, so that much of the material for her book is from her own experience.

This attitude of Garber's is rather typical of the ambivalence we see with regard to property, where it has an immense symbolic significance, whilst at the same time being a place that fulfils an existential need. Nowhere is this ambivalence more evident than in the generic name given to domestic design magazines in the USA. As Garber reports, they are referred to as *shelter magazines*, a designation which carries with it the notion of an imperative condition that could not be further from the concerns of these magazines, with their emphasis on contemporary design and articles on luxury dwellings owned by the rich and famous. Indeed for many people in the UK the term 'shelter' is now so closely associated with the most high-profile homeless charity as to make the use of the term in the USA seem offensive: what could be more inappropriate than the peddling of luxury on a existential condition?

But this improper use of the term 'shelter' is also ironic, in that it demonstrates the act of forgetting that is at the heart of the cult of domestic design. These magazines, which concentrate on the ephemera of dwelling, bring to our attention, as it were by exception, the elemental purpose of dwelling itself: even as we seek to decorate the dwelling and then make it appear modern or contemporary, we still need it to fulfil the same function as the most basic vernacular shelter. The dwelling of M. Hulot can be as effective in terms of shelter and protection from the elements as that of the Arpels, as well as offering so much else that is missing from the modern designed dwelling. The irony, therefore, is that the inappropriate use of the term 'shelter' reminds us of the distinction between appearance and reality, between what we need in order to live, and what we overlay onto this existential condition and then convince ourselves is necessary to our very existence.

I want to pursue this ambivalent attitude to design a bit further by considering what might be called the psychology of the 'make-over': what are the implications of this desire for a fundamental transformation of the place where we live? This will involve looking at what I consider to be so wrong about the notion of design and why it impinges so on the ordinary experience of dwelling. For after all, who is being harmed by the notion of design? Critics like Hal Foster (2002) might indulge in a diatribe against the apparent ubiquity of design, but for those of us not driven by the Marxist imperative, why should it be such a problem? I want to suggest that whilst design might not kill us, it does have implications for our sense of self and the manner in which we relate to dwelling environments. As I have stated at the beginning of this chapter and throughout this book, the problem centres on the notion of acceptance, and whether we can find some true accommodation for ourselves. What design fails to do, as Tati shows in his gentle critique of the bourgeois striving of the Arpels, is to offer us any authentic sense of place, where we are rooted and have clear ruts cut into the fabric that surrounds us.

But first, there is a need to distinguish between the forms of design I see as a problem and the natural actions of keeping our dwelling clean, tidy and as we would wish. The problem is not that one wishes to get the dwelling to a particular standard – a standard that others might approve of or see as tasteless or kitsch – but that design is seen as the end purpose itself. We wish to decorate our properties and to make them pleasant and comfortable environments. But this is because we wish to spend a lot of time in them, and we therefore seek to create places that are convenient and to our taste. We do not see this as involving a fundamental transformation of the dwelling, which would involve the creation of 'standout features' that enhance the financial value of the dwelling or that relate to any particular style. These places are to be *used* not fashioned.

The idea that is peddled in property-based television programmes and in magazines is that the dwelling only becomes entire once it has been designed. It is as if the dwelling can only become fit for purpose once it meets some external standard of design. This reference to the external is of fundamental importance: the crucial problem of design is that it depends on some form of external adjudication or standard-setting through which the property owner can measure what needs to be done to the property. The penchant for the 'make-over', then, leads to the confused idea that housing is a 'property' (King, 2004b) and so it becomes a mere vehicle for design.

There is a similarity here to the manner in which the use of the term 'home' has been appropriated by professionals, such that developers no longer build houses and social landlords do not manage, let and manage dwellings: instead they deal in 'homes'. The effect of this is to diminish the concept, emptying it of much of its significance to us as individuals living with loved ones in our own private place (King, 2004b). The same can be said for the manner in which the domestic 'make-over' has entered the popular consciousness through television, magazines and newspaper supplements. In order for us to redesign our homes we seek the advice of experts, usually flamboyant, larger-than-life characters who seek to impose a particular taste onto their subjects. The effect of this is to professionalise the act of dwelling, by giving undue deference to these experts as the arbiters of taste, and seeking to copy their advice and follow their strictures. But the effect of this is to produce standardised patterns of design based on these 'make-over' programmes. This result is the homogenisation of style, as we put decking into our gardens, and ensure each room has its 'feature' with the appropriate colour scheme to express its mood.

There is, of course, a paradox here: we seek to 'makeover' our own personal space, but on the basis of standardised patterns, following the advice of experts who have never entered the space being 'madeover' except through the anonymity of the plasma screen. This paradox raises the question of whether the 'made-over' dwelling stops being ours in the same manner as it was before? Is it as 'representative' of us as it once was? Perhaps we may feel that it is now more so, as it is still we who have re-made the property to suit ourselves.

But does not the fact that it has been 'madeover' on the basis of someone else's prescription detract from any sense of a personal creation? And is it not rather the case that we feel we cannot use our dwelling fully for fear of detracting from the design? Now, just like the Arpels' we may not see this as a problem. Our respect for the design is such that we are quite prepared to moderate our behaviour, and that of our guests, to fit to the prescriptions of the design. We might now feel that living in the channels created by the design *is* how we ought to live, that we have (re)made our lifestyle through the design. David Fincher satirises this wonderfully in his film, *Fight Club* (1999), with his hero changing his mentality from that of an arch-consumerist, collecting designer products as if they validate his existence, to a maker of expensive designer soap from human fat stolen from liposuction clinics (which is then sold to exclusive department stores frequented by the wealthy and overweight). The hero, played by Edward Norton, implicitly realises the false nature of his past life, and destroys this by blowing up his apartment. He then lives in a derelict house, with a flooded basement, dodgy woodwork and piping. He sets about organising his fight club, which soon develops into a terrorist organisation determined to destroy key commercial and financial institutions.

But what is particularly fascinating about Fincher's vision is how he shows this psychological change. The character's internal changes are shown by a split personality: what he would like to be is shown through an alter ego, whom for most of the film we think of as a separate person called Tyler Durden (played by Brad Pitt). The hero does not know himself or see himself as he is, and he is not aware of the effects he is having on others or himself. Can we not see this as demonstrative of how we ourselves actually do act? We do not tend to acknowledge how we act, but rather we seek to blame changes in our lives and psychology on some external effect. We like to have responsibility externalised in something, be it another person with authority over us (and whom we fear or admire), or the government, or some expert. We want to be told what we should do with and for ourselves, and therefore be able to vest the responsibility away from ourselves. We need our sense of style validated by some outside authority. Thus the radical lifestyle is Durden's, just as all the key decisions are seen as his: Edward Norton's character merely follows along, somewhat bemusedly and apparently unaware of the implications.

However, the sense of publicity goes beyond the displacement of responsibility onto the expert and their standards. It also affects the manner in which we use the dwelling. As with the Arpels, it assumes that the purpose of space is to be admired, shown off and decorated, rather than to be lived in. It therefore inflates appearance above use: we are to admire the decoration and features and ignore the discomfort. We are invited not to judge the dwelling on its use value to the occupiers, but to admire it for its design. Like the Arpels, the host of the designed dwelling is enthusiastic, and keen to point out what they have done, why they did it, and what they still intend to do. They pick out particular facets and relate these to what is important to them in their home: that the house is 'all-communicating'. There is a very real pride in this, a sense of achievement. But we also note that it tends to be rehearsed, having become practised in the saying, so that it becomes a formula, a family story reinforced by partner and children to ensure all parts of the story are told and the full significance given.

There is a difference here between appearance and reality, and we should distinguish between the claims made for design and technology and the reality of the ordinary. Tati's *Mon Oncle* is again so useful to this discussion, as it offers a wonderful example of this sense of delusion, and shows what is lacking in design that prevents the Arpels from fully experiencing where they live. Their house can only be used in a particular manner, according to its strict design principles. Accordingly, the individuals within it are constrained, rather than the house being a source of possibility that allows them to fulfil their ends. Of course, the very aim of design is precisely to increase the possibilities for the users, and to take the house beyond what it currently is and to realise its potential. Yet as Tati shows, this is what it singularly fails to do.

What this shows, and why it is important to this discussion, is a *lack of acceptance*. It is where we are not prepared to accept what we have, and hence we try to design something new out of our existing environment. We feel that, because the dwelling is ours and is therefore special to us, we should make it distinctive and special *in itself*. Yet what is actually special about this dwelling is that it contains us and our own, and this is sufficient in itself.

It is my view that we are rather better at accepting than we think we are, and this means that the effects of any fundamental redesign are likely to fail, if the aim is to create something extraordinary. It is my contention that we are quite able to resist the threats of transformation, and this is precisely why we are much more likely to be watching others transform their property on television than doing it ourselves. We may find the programme entertaining and it might make us wish to change our dwelling, yet for most of us, things do not progress beyond that rather vague desire.

But even where we do undertake a design of our dwelling, we are able to accommodate it without it fundamentally altering what we do, and this shows why we find Tati's description of modern design so amusing. We see it for the futility that it is, and we know that once the designer had left we would very quickly start to cut corners, change things to make them suit and generally fail to live up to (or, better, live with) the principles of the design. The design might alter or moderate our behaviour and habits somewhat, but it is more likely that our behaviour will supplant or change the design. We will start to rub off the corners that stick out, and chip away at the bits that stop us living as we would like.

What this suggests is that the dwelling is still ordinary. It is not, in any intrinsic sense, special or unusual: or rather, it quickly stops being so. As we use the dwelling and move it around to suit our purposes, we begin to take it for granted and do not see it as distinctive, but as the normal and habitual environment in which we undertake our interests and engage with those we love and care for. We find, therefore, that we have not left our ordinary environment. I would suggest this is not because we have moved from one ordinary to another, but that the sense of the ordinary, as it were, *overpowers* any sense of external design.

Of course, this sense of the ordinary is precisely what the 'make-over' seeks to prevent us from accepting. The aim is rather to create a sense of distinction, of difference, so that the dwelling can be considered extraordinary. This yearning creates a churning, or a desire for continual change. It is what we might see as a respectable sense of dissatisfaction: a need for constant improvement and for novelty, but one which is by no means disturbing to one's own sense of probity and which does not make one truly exceptional. This lack of exception arises because we are doing what is popular: we are interested in design precisely because everyone else is, and to be left out would cause us concern.

It is this fact – of our seeking the extraordinary because everyone apparently is – that points to the mismatch between appearance and reality at the heart of the cult of design. What it shows, I would suggest, is that the desire for a 'make-over' of our property contradicts our actual experience of dwelling. We watch these programmes and read the newspaper supplements instead of decorating. We do so in our own private space, which we guard against unwanted intrusion, and where we would not dream of allowing others to dictate on décor and design. We would simply not tolerate anyone's criticism of our taste, even as we watch experts doing precisely that to some naïve family. So our practice of watching in the privacy of our own dwelling is the very antithesis of the publicising of dwelling through a 'make-over'.

This is not to suggest that the ordinary is static, or that to accept what we have is to stagnate. Indeed the opposite is true: *acceptance means modification.* The ordinary is not static: some roots will grow while others wither and die away; entirely new roots take and new plants grow; ruts widen with use and new ruts form in different directions; and so too ruts can become overgrown and eventually disappear with neglect. What the ordinary involves therefore is evolution. When we come to accept the ordinary for what it is, we too sense that it is a dynamic relation, where we will need to change in order to ensure we can fulfil our ends: we change to preserve what we need in our lives to keep them meaningful and fulfilling.

What is important is that acceptance of the ordinary involves the controlling of change. Change, as in nature, is often imperceptible: it is gradual, incremental and not brutal or dramatic. The very stability of our environments tends to prevent this. What this means is that we can accommodate change and do not have to remake ourselves. There is no need for a complete redesign of how we live. It is indeed futile, for it disregards the manner in which we live in the midst of the ordinary, as beings who are capable of modifying their surroundings, but in a way that does not endanger the survival of the aims which dwelling environments are meant to serve. I wish to consider this in more detail by discussing the way in which we use technology and seek to accommodate it into our ordinary lives.

The Ordinariness of Technology

The media theorist Geert Lovink, in his book *Dark Fiber* (2002), considers the manner in which the Internet has developed. In particular, he considers the manner in which the Internet changed from a series of radical, decentred and autonomous networks into what he now sees as just another manifestation of global capitalism. This new medium was to be the base for subversive networks, which connected us but was apart from the globalising tendencies of capitalism. But this has proved to be a case of misplaced optimism, and in *Dark Fiber*, Lovink shows his disillusionment and disenchantment with the commercialisation of the web. He now sees it as a huge missed opportunity, in which the possibilities for limitless, decentred networking have been subverted by the same hegemonic power structures. The Internet has proved to be no different from any other development under capitalism: whatever its potential it soon came to serve the interests of capital, and fell into the hands of speculators and large transnational corporations.

Whilst we can easily write off Lovink's critique as merely another 'no logo' diatribe against global capitalism, what I wish to consider here is another implicit assumption underpinning his position. In seeing the possibility for 'new media', indeed in suggesting some distinction between 'new' and 'old' media, Lovink is falling into the trap of conflating media with message. In this regard he is making exactly the same mistake as those who speculated, and lost, on the millennium dotcom bubble. Just like those who saw Amazon.com and Lastminute.com as opening up a new commercial paradigm, so Lovink has gambled that the means of delivery are more important than the product. The dotcom bubble burst when sufficient people began to realise that these companies were still selling books, gifts and flowers just like traditional companies, and what mattered was the quality and predictability of the service and not the medium. Likewise, the question that immediately occurs to the sceptical reader of Lovink's book is just what were these networks for? One cannot help thinking that the purpose was first to have a network and then think of something to say, and what we see from Lovink's rather tedious histories of these networks, is that much of the discussion was over the purpose of the network and arguments over rules and protocols to govern them.

What Lovink is also unable to appreciate is that the manner in which the Internet developed from an elitist series of networks to a consumer-driven communications web is not a case of capture by global capitalism, but rather a neat example of the development of capitalism itself. As thinkers such as Friedrich Hayek (1978, 1982) and Joseph Schumpeter (1950) have shown, capitalism develops through innovation, risk, and subversion of norms, but in time becomes regularised by laws and common practices. Lovink saw the Internet as a means of subverting capitalism, it being diffuse, uncontrolled and uncontrollable. But this is precisely to misunderstand the nature of capitalism, which too has no centre and develops in an apparently haphazard manner based on revealed preference and the collation of individual choices.

The Internet, being a tool, developed as it did because it was found to have some inherent utility. It could be put to a particular use. What was new was not the uses to which it was put, but the manner in which utility could be expressed. The Internet did not create the expression, it merely altered the means by which it could be conducted. So to extend Robert Nozick's famous dictum from *Anarchy, State and Utopia* (1974), it allowed capitalist acts between consenting adults to be conducted *in a different manner*. What the case of the Internet shows is that first, we need to be careful about conflating the media with the message; but second, it also tells us about the manner in which we use technology. What I wish to suggest is that we accommodate technology into our sense of the ordinary, rather than it being the case that technology shifts us into a new paradigm.

And there is certainly a lot of new technology for us to play with. We are told, particularly by those who aim to sell them to us, that the growing use of Playstations, i-pods and, most particularly, mobile phones and the Internet-connected PC are changing the way we live our lives. We email but do not write letters; we text and lose the ability (or need) to write proper sentences in an established language; we block up our ears with headphones and isolate ourselves from others around us. We do not look at the world around us any more, but instead stare at our mobile phones, anxious we have not missed anything, or text back a response, only occasionally looking up to apologise to or glare at those we have bumped into.

But all the time we are using these technologies, we are assimilating them into our lives and into our sense of the ordinary. Whilst these new technologies alter our behaviour to an extent, it is also the case that we integrate them into our norms and habits, and they become so popular precisely because of their utility and the uses to which we are able to mould them to ourselves. They soon become normal and act as tools or equipment for us to seek our ends. We are in a process of gradual and continual assimilation of new technologies, norms and patterns. William Mitchell argues in *Me++* (2003): 'I assume that we shape our technologies, then our technologies shape us, in ongoing cycles that produce our everyday physical and social environments' (p. 6). Technology is a response to needs and possibilities initiated by us, which then alters our behaviour.

Yet in no real sense does this use of technology become a dependency. We are able to take or leave technologies, even as we may desire them. We do not need to own or use a mobile phone, an i-pod or even a television. On holiday we can go for days away from our email and apparently survive. But we also know that the messages will be there waiting for us, and that we shall have to answer them at some point. Of course, the beauty of email as a tool is that we can answer it at our convenience: it operates according to our time. It is not lost if we are away, and therefore an immediate response is not always necessary. We benefit from the immediacy of the phone line, but without being chased by it.

The Internet is an example of what Mitchell refers to as non-contiguous networks. Certain technologies now mean that networks no longer need to be spatially contiguous. They no longer need to be based on face-to-face relationships, but can be diffused across cyberspace. Our email address book is one such network. Mitchell sees that this has implications in terms of the design of space and the nature of relationships. As he argues, networks are now much less dense, and this implies that relationships are looser and less centred on the individual subject. We can also suggest that the technology takes on some of the responsibility for maintaining relationships, in that they depend upon personalities created in cyberspace: the use of the alias is an important element in displacing identity through these new networks.

However, I would want to diverge somewhat from Mitchell's argument. Whilst email and web chat rooms may allow us to hide our identity, the relation is still centred on the person who maintains the account and address book. The fallacy here is the same as that of Lovink, but at a different level. The fault here is in seeing the network as the main element of the process rather than as merely the conduit between different individual identities. As critical realists such as Margaret Archer (2000) have argued, the problem with social constructivist discourse is that it conflates the means of discourse with the discourse itself. This does not mean that the technology has no effect on behaviour. Indeed the technology dramatically alters the possibilities open to us. But this should not lead us to conclude that the technology is in any way deterministic. Any new tool can open us up to new possibilities, but what matters is the means by which we relate to it.

Perhaps what is significant over the last few years is the particular manner in which technology has become so personalised. As Mitchell has argued, technology is moving from being part of architecture, where things that are located in a particular place, such as the landline phone, PC, television and kitchen equipment, to being placed on the body, such as the i-pod, mobile phone and laptop. This alters the meaning and sense we have of space, in that it becomes more mental than physical, and more technological than architectural. The meaning of the space becomes centred on the technology – the mobile phone or the laptop – rather than the space in which it occurs, be it the office or the dwelling. Accordingly, critics like David Morley (2000) have argued that these technologies are privatising, in that they centre us onto ourselves and allow us to undertake activities as separate individuals rather than in units such as the family. These technologies lead to the fragmentation of the family and work unit because we have personal rather than communal technologies.

But we can see these technologies in a rather different light. Whilst we can appear hermetically sealed within the bubble created by our technologies, they can also allow us to undertake our personal activities much more readily in public. We can take our activities out into public space and so we are now able to work on the train, contact the office or home, listen to our music, and so on. These technologies now centre networks onto us – via the address books we have within our laptops and phones – rather than to a particular space such as the home and the office. We might argue, then, that these technologies actually allow us to socialise practices that have hitherto been restricted to specific locations.

What this suggests, as Mitchell has argued, is that there is a symbiosis between technology and us. It alters the way in which we behave, but we use it also for our purposes and to enhance our activities and the pursuit of our interests. The issue, then, is just how quickly we convert technological innovation into the ordinary. I want to argue that technologies become ubiquitous because of their utility to us. This may alter our behaviour, but in a manner that enhances our ordinary experience. This shows that we are able to control these technologies. I want to consider this by looking at the notion of film and how we now view it.

Film in the Home

The introduction of new technology, in particular the ready availability of DVD technology, has altered the manner in which we view film. Instead of seeing the film in a cinema, as an event with a particular scale, we can now see the film on DVD again and again, we can repeat parts, skip forward, and stop it where we wish. Interestingly this has not meant that we stop going to the cinema; indeed we now tend to watch a film on its initial release and then buy or rent it when it is released on DVD.

But video and DVD technologies serve a further purpose, in that they have made the whole canon of film more available. In the past if we wished to see old films we were dependent on television schedules or film festivals. But DVD technology now allows for the development of an accessible canon to be passed on in the same way as in classical literature and music. We might say that film can now sit alongside these other art forms in having the canon readily available to instant access: we are now very likely to have a library that includes films as well as books.

Stanley Cavell, in his book *The World Viewed* (1979), discusses the ontology of film, being particularly concerned with the manner in which film relates to reality. However, in the preface to the enlarged edition of the book Cavell admits that his remembering of films such as Jean Renoir's *La Régle du Jeu* (1939) was faulty. Writing his book in the late 1960s, Cavell states that he has had to rely on his memory of films, some of which he has not seen for many years. Of course, this was not because Cavell was lazy or not being diligent, but because this was the norm: unless a particular film was being shown in a cinema near you, or you could afford your own personal cinema, there was no alternative to memory.

However, we now no longer have to rely on memory, but can see a film again and again to suit ourselves. This also means that we might not see the filmic experience in the same way as Cavell does. It ceases to be an event in the same way as he understood it. Cavell spoke of his films as part of his youth and development into adulthood: they were significant elements in his cultural growth. They were events in which the viewer had to take themselves out, and into a particular space, and watch in conditions determined by others. But we can analyse a film, and watch it as we please. The film can be ours, in that we can own a copy of it. Moreover, we can now watch our favourite films without having to share the experience and, instead of relying on our one and only view (or at best an infrequent sighting), we have scene selection, instant playback, pause and rewind. Socially, therefore, we experience film differently, and we do not now necessarily experience it together, but watch in private and apart. Cavell was aware of this, when he indicated that a revolution was about to take place in the introduction to his *Pursuits of Happiness* (1981), his study of classic Hollywood comedies of marriage.[1]

This change from communal to personal viewing alters our expectations and probably our attentiveness. Knowing we have permanent access means we do not have to pay the same attention because we can see the film any time. The experience is more ephemeral as we do not have to make the effort to go out: we are not in Cavell's position of having only one or a limited number of chances to see it. Cavell in *Must We Mean What We Say?* (2002, p. 201) talks of a 'phonographic culture', which appears to recognise that the purpose of music has changed from one of performance to that of background. Music is no longer so commonly heard live at a concert, but is recorded, and it is therefore possible to pipe it anywhere. Of course, we can now carry music around on our persons, having downloaded it onto small personal machines such as the i-pod. Music listening need no longer be an intentional act, where we listen attentively. Instead it is 'where music is for dreaming, or for kissing, or for taking a shower, or having our teeth drilled ...' (p. 201).

Cavell seems to recognise a change in the media of perception and I think we can relate this back to the issue of how we relate to film. Technology has altered the medium of perception and this creates differences in intentionality and attentiveness, as well as the mode of possession. Through CDs, DVDs and downloads we can possess film and music in a different manner from Cavell in his discussion of memory and recall in *A World Viewed*. We can now possess it both internally and externally, whereas Cavell could only have internal possession. This makes a significant difference to the manner in which we can recall and review our experiences of film. In a sense, we have a completely different set of sense data: instead of (or as well as) human memory and recall we now have technological recall. Cavell had to rely only on internal possession through the recall of memory and the attentiveness of his gaze. Now we can possess externally through personal access to domestic technologies. We need less attention, as recall can now be mechanical and instantly at hand. At worst, we merely have to watch the film again, and this can be done at our convenience.

What this suggests is that we have come to terms with a particular technology and used it to expand our possibilities. This has not changed us – we might have the same interests and tastes – but we are now more capable of accessing our interests. Those of us who like watching films need not venture out of our own dwelling, nor do we have to sit in a darkened room with people we do not know.[2] This is a subtle process of assimilation: subtle in the sense that we have no plan, but merely an interest to fulfil and now the means to achieve it. So we can add to our ordinary world and extend that set of relations that we are in the midst of.

---

[1] Likewise the increased availability of portable DVD players may well alter viewing patterns just as portable music players have. Hence, in time, the discussion in the previous section on socialising technologies might apply to DVDs as well.

[2] And we also now have the facility with online shopping to have the films delivered to our door. So we truly have no need to venture out!

We may watch in the privacy of our own dwelling, but films still introduce us to the fantastical, and this applies even if we need not pay such rapt attention as we do in the cinema. This is because we are able to control the process through the technology that is available to us. Films introduce the fantastical and magical into our domestic surroundings, at least for as long as the movie lasts and we think about it afterwards. Instead of our going out to see the far-fetched, the tragic, the grotesque, the amazing and the barely believable, it is now there in front of us, spewing from videotape or whirling off a DVD. We can bring the magic into our own home. Yet we can contain it, and localise it into a time and space that we have created. The fantasy is now framed by our ordinary domestic environment. This sense of control is important, and can be shown with an admittedly banal example. My youngest daughter would tend only to watch part of a video attentively and then go off and play, only turning to the screen from time to time. But she would only do this with a film she had seen before and which she was therefore familiar with. However, if we were to ask her if she still wished to have the video on, she was adamant she was watching it. After a while it occurred to us that indeed she was watching it, even as she was involved in some complicated game with her toys. For her, to have a familiar story on was a comfort to her in the same way as we might put music on whilst reading, or have the radio on whilst working in the kitchen. This example tends to confirm Cavell's view that there is indeed a reduction in attentiveness. But it also suggests that we are in control of the use of the technology, and there is thus a significant difference from the 'piped' phonographic experience commented on by Cavell. There is perhaps less of an event in the way we habitually watch film at home – my daughter is usually rapt when she sees a film in the cinema – but we are able to tailor the experience to suit our needs. We can, then, afford a more blasé attitude than if we are visiting a cinema; than if we have gone out, queued to pay our money, bought the popcorn, sat restlessly and with good-natured impatience through the adverts and forthcoming attractions and finally settled down as the titles begin to roll. A video we might watch over several nights after putting the children to bed, paying the bills, doing the washing up and having asked each other what sort of a day it has been. At home we can turn off when the children disturb us because they cannot sleep, or we want a cup of tea.

The importance of this discussion is to show the resilience of the ordinary: we are able to integrate new technology into our ordinary experience. This extends the ordinary and so enhances us. We may find initially that this new technology is extraordinary, and this is precisely because of the possibilities it has opened up. Yet quite soon, we find that these possibilities become probabilities and then certainties. They become a normal part of the life we lead and thus we do not see that we have stepped out of the ordinary. What this shows is that the ordinary is indeed an epistemic condition, a way in which we engage with the world, use it and assert our meanings.

So, to conclude this discussion, what the penchant for design shows us is what, and how much, we have forgotten. We have neglected what is the real purpose of a dwelling: to act as the background to our lives, as the commonplace. If we continually seek to improve what we have, we forget to live now. But our discussion of technology has shown us that it is actually quite safe to forget: the consequences of our neglect are rarely fatal and this is because of the very solidity of dwelling itself. Our dwelling environment is so stable and secure that it can bear some rough treatment. We are able to assimilate much into our ordinary environment even as we seek to change it. This suggests that we are involved in a process of evolution rather than creation, and that we can accommodate change to suit us. This is because, for most of us, the dwelling is a means to the fulfilment of our ends and not the end in itself. We might forget this when we are entranced by programmes on design. But when we turn over to something else we find we are still comfortable and content, and our complex and valuable property slips back into the background where it belongs.

Conclusions

The ordinary, I have suggested, has no beginning and no end, and so I am cautious about being too definitive here. The walk I have taken around the ordinary may or may not have brought me full circle. What I will claim, however, is to have said nothing that was not already in plain sight, if only we had looked. So what I wish to do is make some brief concluding comments, but without committing myself to any ending for the ordinary.

Housing, I have shown and wanted to commit to, is both the root and the rut of our ordinary experience of the world. Like a root, housing provides the foundation to our lives. This goes beyond the mere physical, giving a grounding to our very being and our sense of self. Housing grounds us: it keeps us earthed and settled into something concrete. Just as we form our housing, as we make it as it is, so it helps to constitute us. And finally, it sets us face to face with those we choose to care for and love. We have a confidence in our relations because of what is behind us and supports us. Like a rut, housing provides us with the well-trodden way, with the paths and byways into comfort and security. It guides us through uncertainty, giving us a known course and a clear direction. Housing is known to us like nothing else and so we can act with a security of purpose, in the sure knowledge of our steady progress and our antecedents. We are not the first to pass this way, nor will we be the last.

What we have here is a form of freedom through structure and order, in the sense that Georg Hegel (1991) meant it. It is because of these structures and the sense of certainty they bring to us that we can act in safety and surety. We can set plans and get on with them. We can be what we want to be, or at least make the attempt, knowing we have adequate protection. And this means that the price of failure is small. This is what it means to say that housing is a means and not an end in itself. It provides the physical and ontological structure: it roots us and provides us with ruts, the paths into the world.

The ordinary has a resilience: it is something we can rely on. The ordinary is hard to break. We can treat the ordinary harshly and robustly, and be forgiven. We can rely on its operation and it takes much to stop it, even as we deliberately try with our design and 'make-overs'. This is because it is more than a physical place. The ordinary is what we carry with us. It is formed not just by our current place, but by the memories of places past and the roots they still offer us, and because of the ruts it has made for us into the world. These roots and ruts remain, even if the place is lost. Our roots, so to speak, have an elasticity, so that they can maintain the connection that holds us, that ties us into the earth.

And so we can now see more clearly that the ordinary is a disposition. The ordinary is a way of looking at the world, of taking out of it what we see as significant and important, and of holding on to something solid and secure. I want to end this book, then, by making some more explicit comments on just what sort of disposition this is, and what is the vision I have tried to frame here.

The first point is that this disposition tries to assert the timeless qualities of dwelling and thus what housing creates. Dwelling is always and ever vernacular, and this is not just because, globally speaking, most people build their own dwelling from local materials and always have (Oliver, 2003). Rather what I mean is that if we rely on recall, and on the benefits of hindsight, dwelling is always made by ourselves, using what we have around us. Dwelling is what we are in the midst of, but it has itself come out that selfsame milieu.

This links housing into something broader and into an ontological and epistemological system such that we recognise the emotional and empathetic side of housing as the ordinary. The ordinary comes out of our being in the world, and the disposition it creates is our becoming accustomed to that sense of being. Through an understanding of what it is to be in the midst of the ordinary, we are capable of a new response to housing. As such, the disposition can be seen as an antidote to the sociological readings that predominate in housing theory, with their emphasis on social constructionism, and their relativism. Instead the ordinary speaks of meaning and ideals, and of an anti-materialism that rejects the decentring of the subject and the limiting of agency.

The ordinary is also against the dominance of policy analysis in housing research. It does not see policy as the essence of housing, and it sees no necessary connection between housing and policy. Policy scratches the surface of the ordinary, and merely considers minor aspects that relate to organisations and not to the activity of dwelling itself. We need to separate out the relation between dweller and dwelling, on the one hand, and policy and brick boxes on the other: these are two completely different subjects, and one only begins where the other ends. Policy does not get beyond the front door, but most of what is important to us can only take place once that door is firmly shut. The inability to see this distinction between dwelling and housing policy, and the attempt to conjoin the two together, is destructive of dwelling, even as the policies fail to deliver their intended outcomes or meet their targets.

But to ignore policy begs the question of just what we do look at. If we eschew current developments, the actions of government and statutory and voluntary agencies, what are we to be committed to instead? What I hope I have begun to do in this book is to propose a different form of discourse, one that is impressionistic and linked to memory and emotion, the two predominant elements that cause us to react to our housing as commonplace. This suggests that we look at housing in a particular way, one not bound by policy and the imperatives of production and consumption. Our housing is already around us as the thing that frames us. So to discuss housing is to consider the way in which we experience our day-to-day ordinary lives. Most assuredly, this has implications for the manner in which we build and design our housing. But the point I wish to stress here is that housing is *already here*, both around us and inside us: it is what frames us, yet it is also framed by us. Housing feeds our memory and sense of self, just as it sustains us physically. We need to recognise this and the implications it has for how we view housing, both theoretically and in policy terms. Housing policy can only take us so far – along the path and up to the front door – but after that we are on our own. And this is precisely what we desire: to be on our own, only with those we love and care for: locks are meant to be used.

In terms of a theoretical, or better, a conceptual approach to housing, I hope in this book to have pointed towards a way in which we might further understand this dispositional quality we have towards our housing. Dispositional thought is piecemeal and impressionistic, being sparked by particular events and incidents; we think about things only when we are spurred to do so. And when we have these thoughts they are specific and linked to the context of their firing. The ordinary sense of housing as a disposition is chronotopic, linking us to the specifics of *a* time and *a* place. It is only on the terms of this specificity that we can articulate what housing is for us, as something beyond the merely physical, beyond the made and the planned for. In our writings on housing, then, we should seek to stress this impressionistic quality, where *a* time and *a* place are conjoined. This opens us up to new fields, such as psychoanalysis, Jungian thought and aesthetics, all of which are capable of helping us to see housing differently. These fields are already well-trodden by architectural and cultural theorists, and we can learn much from them to help us.

Yet we already have the most valuable resource of all, and that is our own personal experiences of housing and how we use them. We should not seek to deny the veracity of insights coming from personal experience – we are involved in an activity that is common to all – but rather seek to validate it through the judicious use of theory and the insights of artists such as Bergman, Dreyer and Tarkovsky. We each have our own place fixed in time and we should use this theoretically, just as we use it to ground ourselves in the world.

So whether we choose to see housing theoretically or not – and the ordinary, to state for the final time, is not a theory but a way of looking – what is most important is not merely that we have a dwelling, but what we use if for. What matters is not how dwellings are financed or planned for, but how we personally relate to that particular one which is ours. This relationship centres on our sense of being rooted to that place and of having known ruts made in the ground to help us find our way. And so I conclude where I started, with a root book.

Bibliography

Archer, M. (2000), *Being Human: The Problem of Agency*, Cambridge, Cambridge University Press.

Bachelard, G. (1969), *The Poetics of Space*, Boston, Beacon Books.

Baker, N. (2003), *A Box of Matches*, London, Chatto and Windus.

Bakhtin, M. (1981), *The Dialogic Imagination: Four Essays*, Austin, University of Texas Press.

Benton, T. and Craib, I. (2001), *Philosophy of Social Science: The Philosophical Foundations of Social Thought*, Basingstoke, Palgrave.

Bernstein, J. (2003), 'Aesthetics, Modernism, Literature: Cavell's Transformations of Philosophy', in Eldridge, R. (ed.), *Stanley Cavell*, Cambridge, Cambridge University Press, pp. 107-42.

Birchall, J. (1988), *Building Communities: The Co-Operative Way*, London, Routledge.

Boyne, R. (2000), 'Structuralism', in Turner, B. (ed.), *The Blackwell Companion to Social Theory*, second edition, Oxford, Blackwell, pp. 160-90.

Burke, E. (1999), *Select Works, Volume 2: Reflections on the Revolution in France*, Indianapolis, Liberty Fund.

Burrows, R. (1997a), 'Cyberpunk as Social Theory: William Gibson and the Sociological Imagination', in Westwood, S. and Williams, J. (eds), *Imagining Cities: Scripts, Signs and Memories*, London, Routledge, pp. 235-48.

Burrows, R. (1997b), 'Virtual Culture, Urban Social Polarisation and Social Science Fiction', in Loader, B. (ed.), *The Governance of Cyberspace*, London, Routledge, pp. 38-45.

Cavell, S. (1979), *The World Viewed: Reflections on the Ontology of Film*, enlarged edition, Cambridge, Mass, Harvard University Press.

Cavell, S. (1980), *Senses of Walden*, San Francisco, North Point Press.

Cavell, S. (1981), *Pursuits of Happiness: The Hollywood Comedy of Remarriage*, Cambridge, Mass, Harvard University Press.

Cavell, S. (1988), *In Quest of the Ordinary: Lines of Skepticism and Romanticism*, Chicago, University Press of Chicago.

Cavell, S. (2002), *Must We Mean What We Say?*, updated edition, Cambridge, Cambridge University Press.

Clapham, D. (2002), 'Housing Pathways: A Postmodern Analytical Framework', *Housing Theory and Society*, Vol. 19, no. 2, pp. 57-68.

Danto, A. (1981), *The Transfiguration of the Commonplace: A Philosophy of Art*, Cambridge, Mass, Harvard University Press.

Deleuze, G. and Guattari, F. (1983), *Anti-Oedipus: Capitalism and Schizophrenia*, London, Athlone.

Deleuze, G. and Guattari, F. (1988), *A Thousand Plateaus: Capitalism and Schizophrenia*, London, Athlone.

Dickens, P. (1990), *Urban Sociology: Society, Locality and Human Nature*, London, Harvester Wheatsheaf.

Eldridge, R. (2003), 'Introduction: Between Acknowledgement and Avoidance', in Eldridge, R. (ed.), *Stanley Cavell*, Cambridge, Cambridge University Press, pp. 1-14.

Foster, H. (2002), *Design and Crime and Other Diatribes*, London, Verso.

Garber, M. (2000), *Sex and Real Estate: Why We Love Houses*, New York, Anchor.

Giddens, A. (1991), *Modernity and Self Identity: Self and Society in the Late Modern Age*, Cambridge, Polity.

Giddens, A. (1994), *Beyond Left and Right: The Future of Radical Politics*, Cambridge, Polity.

Gould, T. (2003), 'The Names of Action', in Eldridge, R. (ed.), *Stanley Cavell*, Cambridge, Cambridge University Press, pp. 48-78.

Green, P. (1993), *Andrei Tarkovsky: The Winding Quest*, Basingstoke, Macmillan.

Guerra, T. (2004), 'A Fond Farewell', in Chiaramonte, G. and Tarkovsky, A. (eds), *Instant Light: Tarkovsky Polaroids*, London, Thames and Hudson, pp. 6-9.

Harvey, D. (1989), *The Condition of Postmodernity*, Oxford, Blackwell.

Hayek, F. (1978), *New Studies in Philosophy, Politics and Economics*, London, Routledge.

Hayek, F. (1982), *Law, Legislation and Liberty*, London, Routledge.

Hegel, G. (1991), *Elements of the Philosophy of Right*, Cambridge, Cambridge University Press.

Heidegger, M. (1962), *Being and Time*, Oxford, Blackwell.

Highmore, B. (2002), *Everyday Life and Cultural Theory: An Introduction*, London, Routledge.

Jameson, F. (1991), *Postmodernism, or the Cultural Logic of Late Capitalism*, London, Verso.

Jung, C. (1982), *Aspects of the Feminine*, London, Routledge.

Kemeny, J. (1992), *Housing and Social Theory*, London, Routledge.

King, P. (1996), *The Limits of Housing Policy: A Philosophical Investigation*, London, Middlesex University Press.

King, P. (1998), *Housing, Individuals and the State: The Morality of Government Intervention*, London, Routledge.

King, P. (2000), 'Individuals and Competence', in King, P. and Oxley, M., *Housing: Who Decides?*, Basingstoke, Macmillan, pp. 9-69.

King, P. (2003), *A Social Philosophy of Housing*, Aldershot, Ashgate.

King, P. (2004a), 'The Room to Panic: An Example of Film Criticism and Housing Research', *Housing Theory and Society*, Vol. 21, no. 1, pp. 27-35.

King, P. (2004b), *Private Dwelling: Contemplating the Use of Housing*, Abingdon, Routledge.

Koolhaas, R., Mau, B., Sigler, J. and Werlemann, H. (1998), *S, M, L, XL*, New York, Monacelli Press.

Le Fanu, M. (1987), *The Cinema of Andrei Tarkovsky*, London, British Film Institute.

Lovink, G. (2002), *Dark Fiber: Tracking Critical Internet Culture*, Cambridge, Mass, MIT Press.

Malcolm, D. (1997), *Diana and Nikon: Essays on Photography*, expanded edition, New York, Aperture.

Mau, B. (2000), *Life Style*, London, Phaidon.

Mitchell, W. (2003), *Me++: The Cyborg Self and the Networked City*, Cambridge, Mass, MIT Press.

Moran, J. (2004), 'Housing, Memory and Everyday Life in Contemporary Britain', *Cultural Studies*, Vol. 18, no. 4, pp. 607-27.

Morley, D. (2000), *Home Territories: Media, Mobility and Identity*, London, Routledge.

Norberg-Schulz, C. (1985), *The Concept of Dwelling: On the Way to a Figurative Architecture*, New York, Rizzoli.

Nozick, R. (1974), *Anarchy, State and Utopia*, Oxford, Blackwell.

Oakeshott, M. (1991), *Rationality in Politics and Other Essays*, new and expanded edition, Indianapolis, Liberty Press.

Office of the Deputy Prime Minister (2003), *Sustainable Communities: Building for the Future*, London, ODPM.

Oliver, P. (2003), *Dwellings: The Vernacular House Worldwide*, London, Phaidon.

Packenham, T. (1996), *Meetings with Remarkable Trees*, London, Weidenfeld and Nicholson.

Rosen, S. (2002), *The Elusiveness of the Ordinary: Studies in the Possibility of Philosophy*, New Haven, Yale University Press.

Schumpeter, J. (1950), *Capitalism, Socialism and Democracy*, London, Allen and Unwin.

Scruton, R. (2000), *England: An Elegy*, London, Chatto and Windus.

Scruton, R. (2001), *The Meaning of Conservatism*, third edition, Basingstoke, Palgrave.

Scruton, R. (2004), *News from Somewhere: On Settling*, London, Continuum.

Shiel, M. and Fitzmaurice, T. (eds) (2001), *Cinema and the City: Film and Urban Societies in a Global Context*, Oxford, Blackwell.

Shonfield, K. (2000), *Walls Have Feelings: Architecture, Film and City*, London, Routledge.

Synessios, N. (2001), *Mirror*, London, I. B. Taurus.

Vidler, A. (1994), *The Architectural Uncanny: Essays in the Modern Unhomely*, Cambridge, Mass, MIT Press.

Weil, S. (1952), *The Need for Roots*, London, Ark.

Wittgenstein, L. (1953), *Philosophical Investigations*, Oxford, Blackwell.

Films

Cathy Come Home (1966), directed by Ken Loach.
Damnation (1987), directed by Bela Tarr.
Elephant (2003), directed by Gus Van Sant.
Fight Club (1999), directed by David Fincher.
Gerry (2002), directed by Gus Van Sant.
The Matrix (1999), directed by Andy and Larry Wachowski.
The Mirror (1974), directed by Andrei Tarkovsky.
Mon Oncle (1958), directed by Jacques Tati.
Nostalghia (1983), directed by Andrei Tarkovsky.
Ordet (1954), directed by Carl Theodor Dreyer.
The Others (2001), directed by Alejandro Amenábar.
Panic Room (2001), directed by David Fincher.
Passport to Pimlico (1949), directed by Henry Cornelius.
Persona (1966), directed by Ingmar Bergman.
Poltergeist (1982), directed by Tobe Hooper and Steven Spielberg.
Predator (1987), directed by John McTiernan.
La Régle du Jeu (1939), directed by Jean Renoir.
Repulsion (1965), directed by Roman Polanski.
Rosemary's Baby (1968), directed by Roman Polanski.
The Seventh Seal (1957), directed by Ingmar Bergman.
Solaris (1972), directed by Andrei Tarkovsky.
Spring and Port Wine (1969), directed by Peter Hammond.
Stalker (1979), directed by Andrei Tarkovsky.
Summer with Monika (1953), directed by Ingmar Bergman.
Three Colours: Blue (1993), directed by Krzysztof Kieślowski.
Unbreakable (2000), by M. Night Shyamalan.
The Village (2004), by M. Night Shyamalan.
Werckmeister Harmonies (2000), directed by Bela Tarr.

Index

Printed and bound by CPI Group (UK) Ltd, Croydon, CR0 4YY

22/10/2024

01777626-0020